Lotus

(*Nelumbo nucifera* Gaertn.)

NATIONAL FLOWER OF INDIA

S. C. Sharma
Anil K. Goel
Y. K. Sharma

Bishen Singh Mahendra Pal Singh
23-A, New Connaught Place,
Dehradun - 248 001 (INDIA)
2022

Lotus (*Nelumbo nucifera* Gaertn.) National Flower of India

ISBN: 978-81-211-0733-4

Published by Abhimanyu Gahlot for Messrs Bishen Singh Mahendra Pal Singh, 23-A, New Connaught Place, Dehra Dun, India and Printed at Shiva Offset Press and composed by Doon Phototype Printers, 14, Old Connaught Place, Dehradun, India.

Photographs by: S.C. Sharma, Anil K. Goel and Y.K. Sharma

FOREWORD

Authors Drs. S.C. Sharma and Anil K. Goel were scientists working in the CSIR-National Botanical Research Institute, Lucknow during my tenure as the Director. In 1995 a project on Introduction, Conservation, Documentation, Multiplication and Dissemination of Lotus (*Nelumbo nucifera* Gaertn.) was sponsored by the Department of Environment and Forest, Govt. of India, New Delhi. The objective was not only to set up facilities for collection, study and conservation of Lotus but also to produce a monograph like publication on this important flower. This book is the final outcome of that project.

This book provides detailed information about Lotus, the National Flower of India. It starts with the association of the flower with Hindu culture and then goes on to list its relevance to other ancient civilizations. The introduction describes species closely related to Lotus (*Nelumbo nucifera* Gaertn.) and how different they are from other species. The use of lotus motifs in the architecture has been described giving pictorial evidence. A full chapter has been devoted to the importance of lotus in Hindu mythology followed by its association with Buddhism and in other religions. These religion related chapters have lots of information communicated in an interesting way that I am sure the readers will enjoy. These chapters also bring out how widely the importance of lotus is in the entire South-East Asia, China and Japan.

Blue Lotus that many often confuse with Lotus, is a different genus and species (*Nymphea caerulea* Andr.). The book describes information about its occurrence in various parts of the globe. This is followed by the information on the distribution of genus *Nelumbo*, old records, its presence in different parts of India, Asia, etc. Detailed information on its habitat, its growing requirements in terms of temperature, soil pH, depth of water, etc. has been provided. This is followed by botanical description of its different parts namely the root, the rhizome the leaf, the flower, fruit, seed etc. It is very interesting that the lotus seed viability is more than 1,000 years.

In a subsequent chapter information on its taxonomy, nomenclature, palynology, chromosome number, floral biology and description of the two species of the monogeneric family Nelumbonaceae is provided. Lotus leaf is in a way peculiar because it does not allow water to remain on its leaf surface. The "Lotus Effect" as it is called is due to the biochemical composition of the cuticle that gives its leaf a super-hydrophobic surface characteristic. The authors provide more technical information on this aspect as it has implications for the technology and biology. This information will be interesting for people with a technology bent of mind. The authors have provided greater details about the application of lotus effect in the industrial projects that are being taken up.

Another interesting feature of the lotus is its ability to regulate temperature in its flower that helps insect attraction, pollination and their release after pollination. The authors provide good deal of information on this aspect that will be useful to students of botany. The seed being viable for a very long time, the authors have summarized detailed available information on this aspect.

After briefly discussing the pests and diseases, the information on the cultivation aspects is provided. This information is useful for those who would like to begin cultivation of lotus both for flowers as also for vegetable purposes. Production needs to be followed by its marketing. The authors subsequently dwell upon harvesting, packaging and value for export and the countries that import *Lotus* cultivars. Further more technical details relating to genetic and molecular work has been given including the use of molecular markers in breeding.

Different tissues of lotus plant have medicinal attributes. The book describes the medicinal properties of seeds, flowers, leaves of lotus which is useful for a common man. Further information on such aspects as the chemical composition of nutritionally and economically important parts of lotus and also its perfume has been included.

The final chapters are devoted to how lotus can be used in landscaping, how conservation of lotus could be done, how lotus can be used for bioremediation and potential of lotus as a crop for the economic benefits to the grower. There are several recommendations worthy of following by those involved in the lotus cultivation. I suggest that the Agricultural Universities should take up the R&D work on working out the techno-economics of the lotus cultivation and the progressive farmers should take the advantage of the lotus cultivation in the ponds for enhancing their income. The book is a useful compilation of the work and information on Lotus. The book contains a large number of photographs that enhance the understanding of the written up text besides making the reading more interesting. The information in the book should appeal to both the scientists as well as a common man.

I am glad that Prof. Y. K. Sharma, former Head, Botany Department, University of Lucknow who is a devoted teacher, prolific writer, research scientist, has also been involved in writing this book.

I am sure horticulturists, botanists, teachers, students and those interested in carrying out research on Lotus would find the book very useful. I doubt if so much information on various aspects of Lotus appears in any book or a review. Even for a common man this makes interesting reading. I am sure the book will be received as a reference book and a good addition to the libraries.

P. V. Sane FNA,
Former Director,
CSIR-National Botanical Research Institute, Lucknow

05 April, 2022

MESSAGE

Lotus (*Nelumbo nucifera* Gaertn.) is the National Flower of India and has roots in many Indian religions, particularly in Hinduism and Buddhism, where it is considered as a sacred flower. it is a symbol of purity and eternity in Indian culture. Lotus occupies a unique place in contemporary Indian art, literature and religious rituals. It has great mythological importance indicating its societal relevance even in ancient India. Lotus is one of the oldest flowering plants on earth. In Ancient Egypt, the lotus flower meant creation, rebirth and the sun, while in ancient Greece, the meaning was of innocence, modesty and purity.

All the parts of lotus are useful for humans. The rhizomes and seeds are highly nutritious that has a made lotus a globally sought after commercial crop highlighting its economic importance. Lotus in India represents a great diversity, providing ample scope to improve the crop for specific nutrient traits utilizing its rich germplasm. With advent of several modern tools such as molecular breeding and genome editing, the varieties developed so far utilizing the conventional breeding approach can further be improved to make our varieties globally competitive. Therefore, Indian worker must take up lotus improvement programme in a big way to exploit its commercial potential taking advantage of its rich germplasm and wide agro-climatic conditions, which are suitable for lotus cultivation. CSIR-NBRI has been in the forefront of lotus research in India for several decades and is currently focusing on the trait specific lotus improvement programme besides accelerating conservation actions through large-scale cultivation.

Although research on lotus taxonomy, ecology, reproductive biology, cultivation practices, conservation, and molecular biology has been undertaken world-wide including India for several decades, the results remain scattered, and are mostly unavailable to general readers and lotus growers. This book authored by Dr. S.C. Sharma, Dr. Anil K. Goel and Professor Y.K. Sharma will bridge this gap. I congratulate the authors for bringing almost all the aspects of lotus together that would help the researchers, cultivators, gardeners, students and plant sellers in understanding the socio-religious-scientific aspects of lotus. The book has been written in simple language yet capturing the latest developments in lotus science and has required application orientation to help the entrepreneurs to cultivate lotus as a future commercial crop. I am sure that the book will be well-accepted throughout the world as an excellent reading and reference material.

19/4/2020

Prof. S.K. Barik
Director
CSIR-National Botanical Research Institute, Lucknow

MESSAGE

Nature is powerful, awesome and often beautiful. Flowers are perhaps nature's most subtly attractive features, evolved not only for biological objectives like germination and fertilisation, but touch something deep in human psychology. It is therefore a great pleasure to introduce this beautiful offering on flowers and their varied forms. This book authored by Dr. S.C. Sharma, Dr. Anil K. Goel and Prof. Y.K. Sharma is a beautiful depiction on all aspects on lotus. My close friend Dr S.C Sharma has dedicated his whole life for introducing and studying the ornamental plants especially bougainvillea, gladiolus and rare, endangered and threatened plants. He has laid Japanese Gardens in many CSIR Institutes. Dr. Sharma is a very sincere friend, committed horticulturist and photographer.

I am sure that this book will appeal to the finest sensibilities of readers while educating them of the variety and myriad plants with flowery offerings. Besides the metabolites and other useful phytochemicals, flowers have been appreciated for their mood uplifting and anti depressive properties. From the old and young alike, who can resist the charms of a bouquet of flowers, and break into a smiling and positive frame of mind virtually in an instant? I am sure that this dainty book will bring joy and happiness to all its readers, and also prove to be an asset to the more serious, professional experts as well as the students.

Dr. Girish Sahni FNA,
Former Director General,
Council of Science and Industrial Research,
& Former Secretary, Government of India, New Delhi

MESSAGE

The book written by Dr. S.C. Sharma, Dr. Anil K. Goel and Prof. Y.K. Sharma is a treasure of knowledge on the Lotus, the National Flower of India. It is a very relevant, timely, honest and courageous reminder of the concerns for protecting biodiversity, environment and the future health of humanity as well as the natural and meaningful survival of our world. It brings out very pertinent and important concerns of all alert and thinking human societies. It's not just a scientific statement. This kind of thought and actions are highly relevant for the survival of this world as a whole.

I know Dr. S.C. Sharma for a pretty long time. He is a loving friendly person. Dr. Sharma is a passionate plant collector, who has introduced a large number of rare, endangered plant species in the NBRI, Botanic Garden during his tenure as the custodian of this Garden.

Dr. Sharma has introduced and evolved several *Bougainvillea* varieties. He made significant contributions in the cultivation and marketing of Gladiolus. Today Lucknow is known as the City of Glads. Dr. Sharma is a voracious reader and prolific orator.

For creating education and environmental awareness, Dr. Sharma has established an NGO, Clean and Green Environmental Society for making Lucknow clean and green.

Padma Shri Dr. Mansoor Hasan, MD, FRCP
Founder, LARI Cardiology &
Former Head, Cardiology, KGMU, Lucknow

MESSAGE

Lotus (*Nelumbo Nucifera* Gaertn.) National Flower of India, by Dr S.C. Sharma, Dr. Anil K. Goel and Prof Y.K.Sharma is a brilliantly researched book with fascinating insights regarding this oldest flowering plant on our planet.

This is the first ever book to provide a comprehensive coverage of lotus, the National Flower of India from its extraordinary aesthetic, medicinal and economic values to its significant place in religion, mythology and philosophy.

The authors also share valuable insights on how the lotus has found a prominent place in literature, art and architecture in many parts of the world.

Finally, true to his passion for Botany and Ecology, Dr Sharma's exposition on the lotus as one of the most important aquatic plant species in the Indian subcontinent which is indispensable in its natural ecological environment adds immense value to this impressive magnum opus.

Amrita Dass

Dr Amrita Dass
Founder Director
Institute for Career Studies

Preface

Aquatic plants comprise the most important component of biodiversity on the planet earth and *Nelumbo* (Lotus) has been an integral part of aquatic ecosystem. Here is an attempt to unravel and experience the beauty of the wonderful Lotus, the National Flower of India and the state flower of Haryana, Jammu & Kashmir and Karnataka. The delayed publication of this book has given us an opportunity to incorporate the latest R&D work in the book. We feel great pleasure that this book on lotus shall fill a long time gap of knowledge on its mythological, architectural, philosophical, ornamental, perfumery and scientific aspects with fascinating insights regarding the oldest and relic flowering plant on the planet Earth. There is a paradox in the growth of knowledge. As information accumulates in ever more intimidating quantities, disconnected facts and impenetrable mysteries give way to rational explanations and simplicity emerges from the chaos. Gradually, essential principles of a subject evolve and come into focus. The book is developed alongwith those principles, although there is still a long way to go. The hidden facts provide the excitement and we have tried to arrange them out in a way that will fascinate the readers to connect with lotus, its importance in Hindu mythology followed by its association with Buddhism and other religions, and the importance of lotus in the entire South-East Asia, China and Japan. The lotus is one of the most important wetland plant species in the Indian subcontinent and is 'indispensable to its natural ecological environment'. This adds immense value to this plant species.

The introduction describes species closely related to lotus (*Nelumbo nucifera* Gaertn.) and how different they are from other species. The use of lotus motifs in architecture has been described with colored pictorial evidences. The descriptions also bring out, how widely the importance of lotus is, in the entire South- East Asia, China and Japan. Subsequently, detailed information on its habitat, growing requirements in terms of temperature, soil, pH, depth of water, etc. has been provided. The book includes the taxonomy, nomenclature and palynology, details on morphology, anatomy, shoot, root, leaves and flower with their structural and economic applications. Lotus seed viability of more than 1,000 years makes this taxon of big interest in furthering the paleobotanical investigations. The Flower has the capability of regulating the temperature which benefits the pollination and other floral metabolism. Other than its vegetative propagation, the tissue culture of lotus is also described.

The details on cultivation, harvesting, and medicinal and nutraceutical properties of seeds, flowers, leaves of lotus are also illustrated well. The contents also include its packaging for export and the countries which import and export *Lotus* cultivars.

Among the improvement of lotus quality and quantity not only the traditional breeding techniques but the biotechnological approaches including the molecular markers, are also appended.

This book is principally drafted for teachers, research scholars and students who wish to gain broader scientific knowledge on its history, origin, cultivation, ethnobotany, varieties, and also for the general public, who would like to know more about the lotus. We have attempted to write the book so that even a stranger to science, biology or agriculture may easily pick his way through a vast field of knowledge on Lotus. For this purpose we have provided a much more thorough list of references that even a student or researcher is likely to require. At the same time, we have made an effort, to select mainly those references that are needed, in a reference book on Lotus. We hope that the contents of the book would comfortably satisfy the readers. Greater details about the application of Lotus Effect in industrial projects are also provided.

The authors express gratitude to the Ministry / Department of Environment, Forest and Climate Change, New Delhi for sanctioning and providing financial support for the Project on: 'Introduction, Conservation, Multiplication and Dissemination of Nelumbo nucifera Gaertn.' as well as the Department of Science and Technology, Govt. of India, New Delhi. We express our deepest gratitude to Dr. P. V. Sane, former Director, CSIR-National Botanical Research Institute, Lucknow, who provided all the facilities for undertaking the R and D work on the Lotus. We are grateful to Col. Ajay Gupta, Advisor, Information Technology to the Clean and Green Environmental Society for taking pains for editing the Lotus book. We acknowledge with thanks, the contributions made by Research Associates Drs. Purnima Bajpai and Suresh Singh, for undertaking the plant collection tours and rendering assistance in the R&D work on lotus. If anything is left in this book, we will appreciate if the readers kindly bring to our notice.

Our special thanks to Mrs. Seema Verma for the beautiful handmade Lotus Paintings for the title cover and back cover page.

Thanks are due to Drs. Michel Grant, Amita Kanauzia and K. Radhakrishnan for arranging some photographs.

Last but not the least we profusely thank to our spouses, Mrs. Parvati Sharma, Mrs. Archana Goel and Mrs. Anuradha Sharma for their consistent support and patience during writing this book.

Finally, we express our gratitude to M/S Bishen Singh Mahendra Pal Singh Publishers, Dehradun, for understanding the importance of the publication of the Lotus book.

S.C. Sharma
Anil K. Goel
Y.K. Sharma

CONTENTS

Foreword iii

Messages v

Preface ix

Introduction 01

Chapter- 1 : Lotus in Hindu Mythology 07

Chapter- 2 : Lotus in Buddhism 11

Chapter- 3 : Lotus in other Religions 13

Chapter- 4 : Lotus Philosophy 15

Chapter- 5 : Lotus in Architecture 17

Chapter- 6 : Lotus in Literature 21

Chapter- 7 : Egyptian Blue Lotus 23

Chapter- 8 : Distribution of Lotus 25

Chapter- 9 : Habitat and Ecological Environment 31

Chapter- 10 : Morphological and Anatomical Characters of Lotus 35

Chapter- 11 : Taxonomy and Nomenclature of Lotus 39

Chapter- 12 : Horto-taxonomy of Lotus varieties 43

Chapter- 13 : Floral Biology of the Lotus 47

Chapter- 14 : Lotus Effect 49

Chapter- 15 : Thermo-Regulation of Lotus Flower 53

Chapter- 16 : Lotus Seed Longevity 55

Chapter- 17 : Genetic and Molecular Improvement 57

Chapter- 18 : Conservation of Lotus 63

Chapter- 19 : Propagation of Lotus 67

Chapter- 20 : Pests and Diseases 69

Chapter- 21 : Lotus Cultivation 71

Chapter- 22 : Medicinal Importance of Lotus 75

Chapter- 23 : Lotus Perfume 79

Chapter- 24 : Economic Importance of Lotus 81

Chapter- 25 : List of Lotus varieties 85

Chapter- 26 : Lotus for Bioremediation 95

Chapter- 27 : Lotus in Landscaping 97

Chapter- 28 : Lotus the Crop of Future 99

Chapter- 29 : The Lotus- Immortalized on Stamps and Coins 101

References 103

Lotus

(*Nelumbo nucifera* Gaertn.)

NATIONAL FLOWER OF INDIA

INTRODUCTION

Lotus (*Nelumbo nucifera* Gaertn.) is the National Flower of India and Vietnam. It is the most sacred flower and occupies a unique status in art and mythology from ancient times in India and has remained an auspicious symbol of Indian culture. Lotus is one of the oldest flowering plants on this planet. The flower is perfect, elegant and beautiful. Lotus (Kamal) has remained deeply associated with Hindu culture and mythology. Since time immemorial, the lotus has been the symbol of purity, beauty, divinity and eternity. Sacred lotus is an important constituent of aquatic vegetation as well as aquatic gardening. A pond full of lotus plants in bloom captivates all of us for its beauty. Lotus possesses extraordinary aesthetic, medicinal, nutraceutical and economic values. It has been reported in several Ayurvedic formulations for curing various diseases.

Various variants of Lotus flower in a pond

Lotus is not only important in India but is highly valued in the whole of South East-Asia and world over. *Nelumbo nucifera* 'East Indian Lotus' is an old world Asiatic plant species, now widely distributed in many Asiatic oriental countries, including India, Thailand, Japan, China and Taiwan. It has also become naturalized in North Australia, Hawaii and the Volga River in Russia. This species possesses many racial and ecological variants which have crossed naturally due to its wide phytogeographical distribution, wider adaptation and also by man over the centuries.

A Flower of Faith

The lotus plant possesses versatile symbolic significance for decoration purposes throughout the Indian sub-continent as well as in other eastern countries. The lotus is grown and preserved in ponds and other waterways where the flowers are treated with great reverence. It is synonymous with purity and eternity. The Lotus is found from Kashmir to Kanyakumari showing an enormous phenotypic and genotypic diversity with a large number of racial variants in different shapes, sizes and shades (Goel *et al.*, 2003).

Flower and leaves of Lotus depict India surrounding with various countries and cultures

Lotus flower in various cultures and religions

Another species, *Nelumbo lutea* Pers. 'Yellow Lotus', is a native of North America. Like its other close relative, *N. nucifera* from the old world, this species has also been put to good use as a food crop. The tuberous rhizomes as well as the seeds are used for cooking in different forms. It is also called the New World Lotus. The American Lotus is considered sacred by Native American cultures and is said to have mystic and mythological powers. *N. nucifera* is a highly elegant plant with sweet scented flowers. Typically, the flowers of *N. nucifera* are in various shades of pink or white, distributed in several geographical regions of India. This species has acquired a wealth of several fascinating stories and symbolism, associated with Hindu, Buddhist, Jain and Sikh religions and cultures. It has found a place in our epics 'Ramayana', 'Mahabharat' and 'Bhagwad Gita'.

The lotus flower is said to be the centre of the universe in the Hindu religion. The flower represents our country India and the leaves depict the surrounding cultures and countries. There is a legend that the lotus plant rose from the navel of the Great God 'Vishnu' and in the centre of flower sat 'Brahma'. The role of 'Brahma' was to recreate the universe after the great flood (Goel, 2001).The lotus flower is highly sacred to Buddhists. It is said that Lord Buddha had his birth sitting on the lotus leaf. Buddhism and lotus culture spread together over the whole of South-East Asia (Burkill, 1946, 1965). One of the most pleasing aspects in this taxon is its perennial nature and the rise to faultless beauty from such a miry environment.

The seeds of the lotus are known to possess the maximum period of viability. Scientists have germinated lotus seeds which were 1300 years old. The adult plants flowered well and multiplied (Shen-Miller *et al.*, 1995).

As the maturity in the flowers advances, they tend to stay open longer each day, until the petals are shed revealing the most attractive funnel shaped torus with seeds like a watering can rose. The torus is downy and greenish yellow at first sight and then green, later on drying to a brown woody texture. It may be used particularly as an interesting material for preparing beautiful dry floral arrangements. The nut-like edible seeds are held in pits all over the broad surface of the torus which is the

Lotus temple, New Delhi

Dayal Bagh, Agra with Lotus motif

most important prominent character which separates *Nelumbo* botanically from *Nymphaea* (Heywood, 1978).

Nelumbo (Lotus) has strongly inspired poets, artists and sculptors of both ancient and contemporary times. Lotus motifs have also been frequently used in architecture, especially for decorating the ceilings of temples and forts. During the Mughal period too, lotus motifs were significantly represented in the architecture. The lotus flower was always used as a great symbol of eternal youth. Even many forts of Hindu and Mughal rulers as well as temples were surrounded by lotus covered lakes and ponds. Hampi, situated on the bank of river Tungabhadra, is one of the most important sites in India, included in the World Heritage list. During excavation work a beautiful Lotus Mahal, a fine blend of Indo-Saracenic and Islamic styles of architecture were uncovered by the authorities of the National Museum.

Now a days, lotus has also been given prominence by artists, poets, sculptors, architects and craftsmen. The famous Baha'i temple, also known as the Lotus Temple, at New Delhi symbolizes a perfect lotus flower. The whole structure of this temple looks like a lotus flower in full bloom. In Radhaswami temple at Dayal Bagh, Agra, the marble walls have been carved with exquisite designs of lotus flowers. Lotus motifs are commonly visualized in textile designs and pottery in India, China, Japan and Egypt. It is quite often referred in the folk songs and dances of various traditions, and tribals in India. The offering of lotus blossoms to Gods and Goddesses is depicted through hand gestures in 'Varnam', the introductory sequence of the classical 'Bharatnatyam' dance of South India and similarly in other forms of Indian dances. Lotus flowers are offered to the deities in temples on various occasions during festivals all over the country. Lotus stalks are used to make ornaments, such as necklaces and bracelets in West Bengal. Such ornaments prepared out of lotus stalks have been

Lotus incarving on the front gate of Fatahpur Sikri, Agra

Stalks to make ornaments and stone pendents

mentioned by the great Indian poet, 'Kalidasa', in his most famous play Abhigyana Shankutulam. The lotus has been well decorated in sculptures. Coins, idols in stone, wood and earth have been depicted with the lotus flower. On Indian postage stamps lotus has already appeared significantly. A one rupee silver coin released in 1917 in India by the British, has a lotus flower on top on one side and George V King Emperor on the other side.

During 1857, the first revolt for seeking independence in India, lotus and bread were distributed to spread a message against the British. At the beginning of the twentieth century, as the Indian independence movement seeking freedom from British rule gained ground, the need was felt for a national flag that would serve as a powerful symbol of these aspirations. The first tricolour flag was unfurled on 8th July 1906 during a protest rally against the partition of Bengal, by Sachindra Prasad Bose in Parsi Bagan Square, Calcutta. This flag came to be known as the Calcutta Flag. It had three horizontal bands of equal width with the top being orange, the centre yellow and the bottom green in colour. It had eight half-opened lotus flowers on the top stripe, and a picture of the sun and a crescent moon on the bottom stripe. The words Vande Mataram were inscribed across it in the Devanagari script.

On 22 August 1907, Bhikaji Cama unfurled another tricolour flag in Stuttgart, Germany. This flag had green at the top, saffron in the centre and red at the bottom, the green standing for Islam and the saffron for both Hinduism and Buddhism. The flag had eight lotuses in a line on the green band representing the eight provinces of British India. The words Vande Mataram, in the Devanagari script, were inscribed on the central band. On the lowest band, towards the hoist of the flag was a crescent, and towards the fly a sun. The flag was jointly designed by Bhikaji Cama, Veer Savarkar and Shyamji Krishna Varma.

The National Symbol of India is an adaptation from the pillar in Sarnath near Varanasi (Uttar Pradesh), Lion Capital of the Emperor Ashoka, who ruled from 272 BC to 232 BC.

***Nelumbo nucifera* with matured buds in an aquatic body**

Lotus Effect on a leaf with a water droplet cleaning leaf surface

In the original pillar head there are four lions standing on a lotus flower in full bloom exemplifying the fountain head of life and creative inspiration. Every year on Republic Day, the Government of India awards to distinguished personalities in various fields honours such as 'PadmaShri', 'Padma Bhushan' and 'Padma Vibhushan'. Similarly, the Golden Lotus Trophy is awarded every year to the best feature film produced in India.

A lotus leaf has a characteristic cleaning property. The water repellent lotus leaf surface does not allow any droplet to stick. The dust particles and pathogens are rolled off along with the water droplets keeping the leaf surface entirely clean . This phenomenon is now called as the "Lotus Effect". This property is being applied for developing an exterior paint 'Lotusan' and self-cleaning textiles in Germany. The lotus flower is protogynous and pollination is carried out by insects. The flower works like a thermostat and produces a significant amount of heat during flowering which regulates the temperature precisely. This creates a warm and stable environment for the pollinators, thus facilitating reproduction.

Almost all parts of the lotus plant are edible. The rhizomes, petioles, leaves and flowers are used in savoury dishes. The seeds are eaten raw, roasted or candied. Flowers are in tremendous demand in the floriculture market. The lotus crop has enormous potential for commercial exploitation and profit-earning.

LOTUS IN HINDU MYTHOLOGY

CHAPTER-I

Lotus flowers with other flowers in a plate for offerings at Lord Jagganath temple in Puri, Odisha

शान्ताकारं भुजंगशयनम पद्मनाभं सुरेशं।
विश्वाधारं गगनसदृशं मेघवर्णं शुभांगम।।
लक्ष्मी कांतम कमलनयनं योगिभिर्ध्यानगम्यं।
वंदे विष्णुं भव भय हरम सर्वलोकैकनाथम।।

Salutations to Sri Vishnu, Who has a Serene Appearance, Who Rests on a Serpent (Adisesha), Who has a Lotus on His Navel and Who is the Lord of the Devas, Who Sustains the Universe, Who is Boundless and Infinite like the Sky, Whose Colour is like the Cloud (Bluish) and Who has a Beautiful and Auspicious Body, Who is the Husband of Devi Lakshmi, Whose Eyes are like Lotus and Who is Attainable to the Yogis by Meditation, Salutations to That Vishnu Who Removes the Fear of Worldly Existence and Who is the Lord of All the Lokas.

The lotus flower is deeply associated with Hindu Mythology. for the past thousand of years. The three great Gods of the Hindu religion, 'Brahma', 'Vishnu' and 'Mahesh', with their respective consorts, 'Saraswati', the Goddess of wisdom; 'Lakshmi', the Goddess of good fortune and beauty; and 'Parvati' the 'Shakti' (power) and several other Gods in India are represented posing on a lotus seat. 'Vishnu's idol is always depicted holding a 'Kamal Pushp' (Lotus flower) in one of his four hands. For all female deities, the lotus bud or flower is essential in one of their hands (Goel, 1999). For the Aryans from very remote times the lotus flower appears to have symbolized, first, the idea of a super-human or divine birth and, second, the creative force and immortality.

A painting of Lord Ganesh sitting on the Lotus flower

Goddess Lakshmi sitting on a Lotus Flower

The glorious lotus flower appears to emerge not from the sordid earth but from the surface of the water and is always considered pure and unsullied, no matter how dirty the water of pond or lake may be, thus expressing the idea of supernatural birth and the emergence of the first created object from the primordial water of chaos.

Virtually every God and Goddess in Hindu mythology – 'Brahma', 'Vishnu', 'Shiva', 'Ganesh', 'Rama', 'Krishna', 'Surya', 'Lakshmi', 'Saraswati', 'Parvati', 'Kali', 'Durga' and 'Gayatri' are typically shown sitting on a lotus flower or having a flower in their hands. The lotus thus serving as the seat of deity signifying their divinity and purity is called 'Padmasana' or 'Kamlasana'.

'Surya', the Indian Sun God, is depicted with a lotus in each hand. An open lotus flower looks like the Sun. The lotus flower symbolizes the Sun God because of its ring of ray-like petals around an intensely yellow centre. When the Sun rises, the lotus flower opens its petals as if emitting radiation, and closes when the Sun goes down. Because of this rising and setting phenomenon, it is also the symbol of death and rebirth. 'Saraswati' the Hindu Goddess of learning and music, is usually shown on a swan holding her veena and often has a lotus as her cushion.

'Lakshmi', the wife of God 'Vishnu', is the Goddess of fortune and prosperity as well the epitome of feminine beauty. 'Lakshmi' is portrayed sitting on a lotus, her traditional seat. In Indian mythology Goddess 'Lakshmi' rose from the waters bearing a lotus in her hand. All forms of 'Lakshmi' (Gaja Lakshmi) have elephants offering lotus flowers on each side. The lotus is the

Lord Vishnu on a Lotus Flower

Lotus that grew from the navel of God 'Vishnu'

flower of 'Mahalakashmi' who is the supremely attractive aspect of the Divine Mother's power of beauty and harmony. It is said that 'Shri Krishna' had the imprint of a lotus on the soles of his feet (Mitra & Kapoor, 1976).

The lotus is much more than a sacred flower for Hindus. It is symbolic of all that is good, pure and beautiful. Hindus are infatuated with the lotus flowers. The scriptures tell us that it first bloomed at the creation of the universe. Over the centuries, the lotus flower of paradise has not lost its allure and charm. The lotus stands for beauty, activity, creativity, wealth, health and prosperity. Right from Vedic times the lotus has been the symbol of Mother Earth or the Universe (Goel, 1999).

According to Hindu mythology, after the great deluge the Almighty fertilized the waters to produce an egg. From this came 'Brahma' the creator, and from second egg the universe. Hence, the universe is called 'Brahmand' or 'The Cosmic Egg' in Sanskrit. Another legend states that Lord 'Brahma' emerged from the lotus that grew from the navel of God 'Vishnu' the preserver.

The Goddess of wealth, 'Lakshmi' is called 'Padma' and 'Kamla' after the other names of the lotus flower. The four handed 'Vishnu' is depicted bearing a conch, chakra, mace and a lotus flower in each of his hands. There is the custom of offering 108 lotus blooms for worshipping 'Devi Durga' and 'Laxmi' on the occasions of 'Durga Puja' and 'Deepavali' festivals. The lotus blooms are also offered to Lord 'Shiva' during 'Maha Shivratri'. The streets surrounding the 'Vishwanath' temple in Varanasi and the 'Lingaraj' temple in Bhubaneswar and the

"Padmnabh" temple in Thiruvananthapuram are flooded with florist shops selling pink and white lotus buds and flowers for offering to Lord 'Shiva' and 'Vishnu' during the religious and most auspicious '*Shravan Mas*'.

The lotus is one of the world's most celebrated flowers. From time immemorial to the present day, it has always been in folklore, religions and the arts in one form or the other. Scriptures tell us that the lotus first bloomed at the creation of the universe. Over the centuries lotus has not lost its perfection and attraction. The lotus is extremely special flower to the Hindus, as the foremost symbol of beauty, purity, prosperity, fertility and eternity. According to Hinduism within each human inhabiting the earth is the spirit of the sacred lotus. The lotus is considered to be the doorway to the womb of the universe.

In the 'Bhagavad Gita', man is adjured to be like a lotus. He should work without attachment, dedicating his actions to God, untouched by sins like water on a lotus leaf and the beautiful flower standing high above the mud and water.

According to 'Devi Puran', five flowers are the favourites of Gods—'Lotus' (*Nelumbo nucifera*), 'Kaner' (*Nerium odorum*), 'Champa' (*Michelia champaca*), 'Mango' (*Mangifera indica*) and 'Shami' (*Prosopis cineraria*). Another aquatic plant species which is highly valued in the Indian mythology, religion and culture has been *Euryale ferox* Salisb. (Euryalaceae) commonly found under cultivation in North Bihar region adjacent to Nepal.

A detailed examination of the mystic and artistic records of India and South – East Asia by Mc Donald (2004) have revealed that the famous psychotropic plant species of the ancient Aryan was the sacred eastern Lotus *(Nelumbo nucifera)*. Since most of the Hindu and Buddhist Gods and Goddesses have traced their origin from the Vedas and have always shared very close symbolic associations with *Nelumbo*, there is every reason to believe the divine status of this most sacred plant species deriving from India's prehistoric past dates back to the Aryan civilization (Mitra & Kapoor, 1976).

***Euryale ferox* in a pond with fruits and fresh arillated seeds**

Lotus blooms in a Mithila Painting

With so much appreciation of this sacred plant, there is no exaggeration in saying that the Indian lotus has always enjoyed a high reputation since time immemorial until now, an unparalleled popularity throughout the length and breadth of one great nation of India as well as the whole of South-East Asia.

LOTUS IN BUDDHISM

CHAPTER-2

Lotus garden in Buddhist temple at Bali in Malaysia

The sacred lotus flower is a symbol of enlightenment in Buddhism. It grows in murky water and sprouts to produce beautiful flowers. This fact reminds us that Buddha was born in this dirty world and was enlightened at the end. Buddhist temples are built with sculptures of Buddha sitting or standing on the lotus flower. The sutras say that the lotus flower has four virtues: fragrance, purity, softness and loveliness. Many flowers have these characters but the lotus also possesses other characteristics which set it apart from them and make it symbolic to Buddhism. Lotus plants produce flowers, leaves, as well as fruits at the same time. Buddhists thought that the lotus could reveal past, present and future all at one time.

Buddha sitting on a lotus resembles the great God Brahma, the creator of this Universe. Gautam Buddha was born in 560 BC in Lumbini near the India-Nepal border. When he was born, he walked seven steps in ten directions. With every step he took, a lotus flower appeared.

According to Esoteric Buddhism, the heart of the human being is like an unopened lotus flower. Lotus symbolizes purity; when the virtues of Buddha develop therein, the lotus blossoms. In Buddhism, the lotus flower symbolizes faithfulness. Teachings of Buddhism respect the lotus as the "Holy Flower", so the cultivation of the lotus was popularized in the courtyard gardens of Buddhist temples.

As Buddhism spread from India to Central Asia and China in the first few centuries AD, lotus flowers were used to represent Buddha. They were featured on rosettes, scrolls, motifs and iconography. In Tibetan Buddhism the principal mantra is "Om Mani Padme Hum" This means "Hail, jewel in the Lotus". One can see lotus motifs in famous Buddhist monasteries all over the world.

The Sacred Ashtamangala in Buddhism

The Ashtamangala (refer https://en.wikipedia.org/wiki/Ashtamangala) is a sacred suite of Eight Auspicious Signs endemic to a number of religions such as Hinduism, Jainism, and Buddhism. The symbols or "symbolic attributes" (THL: chaktsen) are yidam and teaching tools. Not only do these attributes (or energetic signatures) point to qualities of enlightened mindstream, but they are the investiture that ornaments these enlightened "qualities" (Sanskrit: guna; THL: yönten).

Many cultural enumerations and variations of the Ashtamangala are extant.

Conch or Shankh
Endless or infinite knot
Pair of Golden Fish
Lotus or Padma (पद्म)
Parasol
Treasure Vase
Dharmachakra or Wheel
Victory Banner

In Buddhism, Groupings of eight auspicious symbols were originally used in India at ceremonies such as an investiture or coronation of a king. An early grouping of symbols included did not include the Lotus flower (padma) but later on were always included. In Buddhism, these eight symbols of good fortune represent the offerings made by the gods to Shakyamuni Buddha immediately after he gained enlightenment. Tibetan Buddhists make use of a particular set of eight auspicious symbols, Ashtamangala, in household and public art. Some common interpretations are given along with each symbol although different teachers may give different interpretations.

The Lotus Flower or Padma

The lotus flower (Sanskrit: Padma; THL: Péma), represent the primordial purity of body, speech, and mind, floating above the muddy waters of attachment and desire. The lotus symbolizes purity and renunciation. Although the lotus has its roots in the mud at the bottom of a pond, its flower lies immaculate above the water. The Buddhist lotus bloom has 4, 8, 16, 24, 32, 64, 100, or 1,000 petals. The same figures can refer to the body's 'internal lotuses', that is to say, its energy centres (chakra). A rare 1000 petal Lotus bloomed at Vazhapally Mahadev temple in Kottayam, Kerala.

A Lotus flower with 1000 petals

LOTUS IN OTHER RELIGIONS

CHAPTER-3

The lotus has something very special about its flower to earn this great honour and recognition in Shri Guru Granth Sahib, the most sacred book in Sikh religion ('Adi Granth Sahib' Gurbani). For example, the Gurbani (SGGS) says-

Enshrine God's Lotus Feet within your heart.
I have enshrined the Lotus feet of the Guru within my heart.
Divine Lotus feet dwell in the minds of his devotees.
God's Lotus feet are the support of the soul.

The ancient Egyptians from the fourth dynasty greatly valued the Sacred Lotus in all their religious ceremonies including the funerals. In Egypt, the blue lotus (*Nymphaea caerulea*) is a symbol of the Sun God. Actually it is not a true lotus. The lotus was introduced to Egypt about 2500 years ago but is no longer found in the Nile region. To the ancient Egyptians, a lotus bud or 'Sesen' was a symbol of rebirth. It closes in the evening and falls to the water, in the morning it reopens and is lifted above the water surface. The behaviour of the lotus emulates that of the sun because of this rising and setting. It is also a symbol of death and rebirth. According to one creation myth, a giant lotus emerged from the watery chaos at the beginning of time. Out of its centre, the sun itself rose on the first morning.

Myanmar is a country dotted with several ponds and lakes, large and small. 'Padon-na-kya', the sacred lotus is believed to bloom only in the sunlight. When younger people pay obeisance to elders, the latter usually respond with blessings "May you be as fresh and bedewed as the Padon-na-kya". The lotus motif is also found as a decorative part in the architecture of Buddhist shrines and sacred depositories such as Chedis. It decorates the pinnacles of tiered roofs of monasteries and palaces. There is also a vessel somewhat like a fruit stand decorated with lotus petals for offering food and fruit at sacred Buddha shrines. The exotic lotus is a motif which also adorns the golden thrones on which Buddha images are placed.

The Thais, ninety percent of whom are Theravada Buddhists, hold the lotus plant in particularly high esteem and use it in a variety of religious and cultural ceremonies. The lotus is associated with heavenly beings. Many Brahman Goddesses have lotus blooms in their hands when portrayed in the paintings. A variety of water lilies share the same Thai name as the lotus.

In China, the lotus symbolizes the Confucian model of the 'Superior man' and is one of the noble flowers. China has hundreds of Lotus gardens and all parts are cooked or eaten by the natives. In Vietnam, there is a saying, "Live in Society without being contaminated by its vices." "To evoke the quality of this man be like the lotus" one often says in Vietnamese religion. Parsi people believe that the open lotus flower is a symbol of the Sun God.

When the followers of Bahaullah, a seventeenth century Iranian Mystic, decided to build a temple in New Delhi, their choice was the lotus. Fariburz Sahaba, the architect who made the design explains, "The lotus not only has an association with all the religions of India but is probably the perfect flower of the whole world. It is symmetrical and exquisitely beautiful. It grows in a swamp and raises its head out of the slime absolutely clean and perfect". Lotus has also been mentioned in Bible. There are Lotus paintings inside the Thumboly Church near bypass, Alappuzha district in Kerala.

A lotus flower lives for only four days. The Japanese describe blossoming in the following way, first day flower has the shape of a bottle of sake (Japanese rice wine) the second day that of a cup of sake, the third day of a soup bowl and the fourth day, that of a saucer.

Thumboly Church near bypass, Alappuzha district having lotus paintings on its walls

A beautiful lotus bloom with pink stripes on petals

LOTUS PHILOSOPHY

CHAPTER-4

Lotus is an excellent, all-purpose remedy for enhancing and harmonizing the higher consciousness, particularly for integrating spirituality in a balanced manner with the other energy centres. The lotus inspires a person whenever he is faced with trouble, so that, no matter how dark the waters may get, he or she will rise above it all let the beautiful flower in them bloom as long as they have a pure and sincere heart. To the mind of ancient worshippers, the lotus was symbolic for the rise of the soul of a man.

The physical characteristics of the plant and the beautiful blossoms reflect its spiritual symbolism and properties. The plant has its roots deep in the mud underwater. The long, cylindrical stalk passes through the water and rises above the water surface. The mud in which the lotus roots grow represents material life, while the water through which the stalk passes represents the astral world. When the plant reaches the surface of the water and opens its bud to the sun, it represents spiritual well being and enlightment.

The crown chakra - the energy centre at the top of the head- is also called 'the thousand petalled lotuses', which is the symbol of final revelation. It is believed that the transformation of the world into paradise can occur through the lotus, which expounds fully the oneness of all life. The fully opened lotus flower has a strong solar character, and its petals are linked to the rays of sun which is a symbol of enlightenment. Meditating on the lotus flower brings harmony into all aspects of our well being, material, mental as well as physical.

A hundred petalled dark pink lotus

The idea of enlightenment is symbolized by the life cycle of the sacred lotus plant because it begins humbly in the mud of ponds but soon grows and sends stems and flowers well above the surface of the water, thus showing the path of spiritual enfoldment.

In 'Bhagavad Gita', man is adjured to be like the lotus. He should work without attachment, dedicating his actions to

God, untouched by sin like the water on a lotus leaf and the beautiful flower standing high above the mud and water.

Lotus flowers inspire people to continue striving through difficulties and to show their best side to the outside world, no matter how difficult the circumstances may be. This is understood as being just like the lotus flower, bringing beauty and light from murky darkness at the bottom of the pond.

Padma-asan: The seated posture in which the legs are crossed and locked is called the lotus pose, 'Padma-asan'. The right foot is placed on the left thigh and the left foot on the right thigh with the soles upward. The hands are placed on the thighs with the palms upwards. With patience and regular practice of Padma-asan, several human diseases can be cured. It is also helpful in spiritual and mental attainment.

Lotus Chakra: The lotus is an excellent all-purpose remedy for enhancing and harmonizing the higher consciousness, especially for integrating spirituality in a balanced manner with the other energy centres.

A hundred petalled pink double race

LOTUS IN ARCHITECTURE

CHAPTER-5

The Lotus in India's History

The motif of the lotus flower has been extensively used in the architecture of palaces, temples and buildings by various kings, rulers and saints throughout all times. Temples, forts, monuments and buildings are well decorated with lotus motifs. Hindu, Buddhist and Jain temple railings are illustrated with the lotus. We find beautiful carvings of the lotus in Somnath, Sanchi and Amravati temples. The Golden Temple (Swarn Mandir) in Amritsar the most important temple of the Sikhs is full of lotus floral motifs. All Gurdwaras are illuminated with the lotus flower. Historical monuments, such as the Red Fort (Delhi), Taj Mahal (Agra) and Charminar (Hyderabad) have been well carved with the lotus. We find beautiful engravings of the lotus in the Radha Swami Temple at Dayalbagh, Agra.

The lotus motif has also been widely used in shrines, art, architecture and sculptures of the Jains. The Jain

Lotus Pond in Taj Mahal, Agra

Lotus carving on stone at Dayalbagh Temple, Agra

Lotus incarvings at Dayalbagh, Agra

Ancient Vishnu & Lakshmi idol holding lotus bud

Lotus motifs in the Radha Swami Temple at Dayalbagh, Agra

A view of Ramkrishna Math, Lucknow building depicting the blooming lotus with a lotus pond in frontage

temples at Dilwara in Mount Abu, Rajasthan, evoke a sense of awe from their sculptured artistry. These temples date back to the eleventh to thirteenth centuries AD. The most outstanding feature of these temples is the thousand petalled lotuses that decorate the domes in the *rangamandapas,* signifying a very highly evolved technical artistry. The statues of Tirthankaras are depicted with lotus flowers in their hands. The design of Sri Ramakrishna Math in Lucknow is in the pattern of a blooming lotus flower which is a fine piece of architecture. The open lotus flower is the symbol of divine origin, purity and spiritual elevation.

The Lotus in Mughal Era

Mughal rulers came to India from Central Asia in the fifteenth century and by the seventeenth century ruled over most of the Indian subcontinents. The first Mughal Emperor, Babar (1483-1530), constructed a lotus tank which represented paradise near every mosque. It also symbolized triumph over darkness. Many beautiful buildings were constructed in the Mughal era using lotus motifs on the walls and roofs. One of the favourite motifs in Mughal architecture was the lotus. Mughals were great lovers of the ancient form of Persian architecture known

Lotus incarving on a wooden panel in Ramkrishna Math, Lucknow depicting lotus petals

Lotus incarving on glass panel in Ramkrishna Math, Lucknow

as 'Pairi-daeza', or the Paradise Garden. The lotus has also been well depicted in Persian architecture. One of the finest examples can be found in the Red Fort, Delhi, in a part of the palace built by Emperor Shah Jahan. Here the lotus appears in the harem. The lotus flower has been cut from a single block of white marble. The beautifully carved lotus flower with its leaves looks extremely attractive.

Baha'i Lotus Temple

One of the key concepts associated with the lotus is light. The flower lends itself to this observation as it opens its petals to the sunlight and closes them to darkness. In all religious traditions, this connection with light evokes not only divine power but also higher levels of illumination of the human mind.

In New Delhi, the National Capital of India, a lotus flower shaped Baha'i temple has been built. It is a rare piece of architecture with high aesthetics, which is a fusion of art, culture and religion. It is appreciated as the Taj Mahal of the twentieth century. The attractive design of the temple creates the visual effect of a large, white lotus blossom. The Baha'i Lotus Temple emerges from a pool of water. It is a metaphor for the physical to spiritual journey. The carefully tended garden and the beauty of its ground create a vision of Paradise on earth. The purity of the white exterior by day and the floodlit form during the night directly associates the Lotus Temple with the power of light and all its transcendent symbolism.

The famous Opera House, icon of Sydney, Australia facing, the sea is also one of the finest examples of modern architecture in the shape of a lotus flower floating on the water.

Front view of Opera House, Sydney, Australia

LOTUS IN LITERATURE

CHAPTER-6

We find references to the lotus (Kamal) in 'Vedas'. In 'Rigveda' it is mentioned that the earth has emerged from water, now confirmed scientifically. The lotus is the symbol of the universe. In 'Atharvaveda' the womb of the mother is compared with the lotus flower.

The lotus has been significantly mentioned in literature in Sanskrit, Hindi and several other Indian regional languages. The beauty and importance of the lotus flower has been widely appreciated throughout all times by the great Indian poets in their literary works. Kalidas has mentioned the lotus in *'Meghdoot'* and *'Abhigyan Shakuntalam'*. Tulsidas has elaborated on the theme of the lotus in *'Ram Charit Manas'* and *'Vinaya Patrika'*. Bihari in *'Bihari Satsai'*, Mirabai, Surdas, Bharthari, Vidyapati, Kabir, Nanakdev, Jayasi, Keshavdas, Bharatendu Harish Chandra, Jagnnathdas Ratnakar, Jaya Shankar Prasad, Surya Kant Tripathi 'Nirala', Sumitra Nandan Pant, Mahadevi Verma, have all greatly appreciated the spectacular beauty of the lotus in their literary works.

Guru Ravindra Nath Tagore has said that the Indian culture is like a hundred petalled lotus in which each petal depicts a regional language and culture. This symbolises perfect unity in diversity and harmony.

The beauty of the lotus flower and its petals has been compared with different body parts of God and Goddess. Most beautiful lady is called *'Padmini'*. The lotus possesses the prime place in Indian life and literature. The lotus eyed and lotus faced heroines are legends. Beautiful hands and feet are always compared to the lotus. Love letters were written on lotus leaves. Eyes of female beauty are compared with the lotus and called *Kamalnayani*. Considering the beauty of the lotus, ladies are given the names Kamla, Padma, Ambuja and Neerja. Lotus has also been described well in the plant myths and traditions in our country (Gupta, 1971).

Comparing the eyes for beauty with Lotus 'Kamalnayini'

In classical dances of India, Japan, China and other Asian countries the lotus is symbolized by the hands in different *'mudras'*. In *'Bharatnatyam'* the lotus is offered to God and Goddess in the form of *'mudras'*. Ancient and modern ornaments and clothes are well decorated with the outline of the lotus plant and flower.

Table-1: Names of the Lotus in Indian Languages

Sanskrit	:	Ambuja, Padma, Pankaj, Kamla
Hindi	:	Kamal, Pundarika (White flower), Kokanad (Red flower) Neerja, Rajiv, Ambuj
Urdu	:	Nilufer
Kashmiri	:	Pamposh
Bengali	:	Padma
Oriya	:	Padam
Assamese	:	Podum
Manipur	:	Thambal
Marathi	:	Kamal
Gujarati	:	Suriya Kamal
Tamil	:	Ambal, Thamarai
Telugu	:	Kalang, Erra-tamara
Kannad	:	Kamala
Malyalam	:	Thamara, Senthamara
Sindhi	:	Pabban
Pali	:	Paduma, Uppala, Pankaja, Pundarika, Ambhoja, Varija, Satapatta, Kamala, Kumuda

Table-2: Names of the Lotus in Foreign Languages

English	:	East Indian Lotus, Lotus, Oriential Lotus, Sacred Lotus
French	:	Lotus Indien
German	:	Indischer Lotus
Italian	:	Fior di Loto, Loto
Spanish	:	Loto Sagrado
Portuguese	:	Flor-de-Lotus, Lotus-da-indica
Chinese	:	Lian Hua
Taiwan	:	Lian
Japanese	:	Hasu
Korean	:	Yon Puri
Thai	:	Bua Luang
Vietnam	:	Hoa Sen, Lien Hoa
Sinhalese	:	Nelum
Myanmar	:	Padon-na-Kya
Nepalese	:	Rato Kamal, Kamal
Persian	:	Nilufer

The botanical treatise, **Hortus Malabaricus** (1678-1693) has been a 12 volume work dealing with 742 plant species documenting the plant wealth of Malabar region by a soldier in the Dutch East India Company: H.A. Van Rheede. Lotus has been illustrated in Vol. 11 of **Hortus Malabaricus** as *Nymphaea nelumbo* (as Tamara) at page 59, table 30.

Tamara (*Nelumbo nucifera*) as *Nymphaea nelumbo* in HA Van. Rheede, Hortus Malabaricus (1678-1693), vol. 11; p.59, t.30.

EGYPTIAN BLUE LOTUS

CHAPTER-7

The blue lotus is not a true lotus (*Nelumbo nucifera* Gaertn.), family Nelumbonaceae. The blue lotus (*Nymphaea caerulea* Andr.) is treated under the family Nymphaeceae (Waterlily family). The blue lotus has several common names including Egyptian lotus, blue water lily and sacred lily of the Nile. Fossils of this plant date back to the Jurassic period, about 160 million years ago. Amazingly, the fossil records suggest that the blue lotus has not changed much in characters. The Ancient Egyptians believed that the Nile Valley area was the birth-place of human civilization. That is why the tributaries of the river Nile are shaped like fallopian tubes, symbolizing the birth of mankind. The blue lotus had a profound impact on human society and civilizations, including the ancient Egyptians, Chinese, Japanese, Indians, Buddhists, Greeks and numerous other cultures around the world.

Nymphaea caerulea Andr. has beautiful blue flowers, 8-15 cm across, held high above the water. The sepals are greenish-white and thickly marked with black lines and dots. The narrow blue petals with pointed tips are blue above and dull white in the lower half. The large rootstocks are eaten in several regions of Africa. This species is not as showy as some of the blue hybrid cultivars. It has never been a prolific bloomer. Its common name of 'Blue Lotus' is misleading as it is not a true lotus.

In India the "Blue Lotus" has been *Nymphaea stellata* having stellate (star like) beautiful flowers with violet undersides and prominent green veins. *Nymphaea capensis* (Cape blue water lily) is a highly popular water lily possessing fragrant blue flowers with green sepals and yellow stamens. The flowers last very well as cut flowers. Some of these water lilies are quite vigorous and display generous blooms for a long time.

Blue Nymphaea in a pond

DISTRIBUTION OF LOTUS

CHAPTER-8

The genus *Nelumbo* Gaertn. distributed over a wide range of phytogeographical regions, exhibits great diversity in shapes, sizes and shades of red, pink and white flowers having 8-1000 petals. This species, generally known as the 'East Indian Lotus', is an old world Asiatic taxon widely distributed in many tropical and subtropical Asiatic countries, including India, Sri Lanka, Myanmar, Indonesia, Malaysia, Thailand, Korea, Vietnam, Japan, China, Russia and Taiwan. It is also naturalized in North-Eastern Australia, Hawaii and in the Volga river delta area where the river flows into the Caspian Sea (near Southern Russia). POWO (2021) cites nativity of *Nelumbo nucifera* as southern China whereas it is native to the Indian sub-continent and described in all ancient literature for thousand of years in India.

Pink Lotus

Genus *Nelumbo* Adans. is widely distributed throughout the tropical and subtropical regions of the world. It is found in lakes, ponds, quiet streams and estuaries of tropical and subtropical zones. The other wild species of this genus is *Nelumbo lutea* Pers. which is distributed in North America, from New York and Southern Ontario, throughout the Eastern States to Florida, Texas to Central America, Columbia and West Indies. It has been growing profusely along the delta of the Mississippi River (Borsch & Barthlot, 1994).

Fossil records of *Nelumbo* have been recorded from the Cretaceous to Pleistocene periods and reported from Southern and Western areas of North America, Greenland, Central and Western Europe, Russia, Northern Africa, Japan and India. Miocene flora of Central Europe included this genus which also occurred in Hungary in the Oligocene period and as far as North Canada during the Eocene period. The flora was redistributed during the Pliocene period in North

Lotus (pink) flowering in a pond in wild

Nelumbo nucifera **(A light pink cultivar)**

America and Europe as the climate steadily deteriorated and started changing fast. The climate became cold and the genus *Nelumbo* and some other hydrophytes became extinct on the first approach of glacial conditions in Europe and parts of North America (Sculthorpe, 1967).

Puri (1950) described the first fossil of *Nelumbo* from Pleistocene deposits of Kashmir in India. Lakhanpal (1955) reported the fossil record of this taxon from the early Tertiary rocks of India. The presence of *Nelumbo nucifera* from Tertiary period in Assam has been recorded on the basis of impressions of leaves and rhizomes discovered in a collection from the Eocene beds near Damalgiri. According to Berry (1917), *Nelumbo* has a long geological history during which ancestral species occupied areas between its present distribution in the wide apart extremities. Berry's opinion has thus been well

A flower of White Lotus

A flower of dark pink double lotus

supported by the above fossil records. These records reveal that *Nelumbo* had been growing in the Indian subcontinent for millions of years, and they provide enough evidence in support of the view that *Nelumbo nucifera* is indigenous to this country. In India, *Nelumbo nucifera* has been widely distributed in the ponds and lakes in almost all the phytogeographical regions of the country ranging from Kashmir to Kanyakumari (Roxburgh, 1874; Coventry, 1927; Biswas & Calder, 1955; Lavania *et al.*, 1990). During intensive survey, it has been revealed that the population of *N. nucifera* is dwindling considerably in the Indian subcontinent. *N. nucifera* is spectacularly seen blooming round the year in ponds on both the sides of the highway from Thiruvananthapuram to Kanyakumari in various shades of pink and white.

In Manipur the lotus grows widely in the wetlands. But in recent years the population of this plant species in the wild habitat is declining due to rapid shrinkage of the wetlands and land use change of water bodies. A small population of lotus is also cultivated in ponds of many households in the valley of Manipur. Lotus is sacred to the people and it is popularly believed that it lowers the level of water pollution. The Sanapat Lake (freshwater lake) is located in the Bishnupur district of Manipur state in North-Eastern India. The name 'Sana' means golden and 'Pat' means lake. The lake spread over 55 hectares is a natural lotus growing area. It was a beautiful lake in the recent past but now it is dying rapidly. It is facing serious threat from various biotic pressures particularly the human activities like artificial lake

A shrinking aquatic body due to various biotic factors

Shrinking aquatic body due to fast urbanization

A degraded aquatic body

Degrading aquatic bodies adjacent to urban localities

modification, marginal paddy cultivation and unscientific exploitation of natural resources. Sanapat Lake has shrunk by 62% of its area within a span of 25 years. The lake is under pressure from its surrounding human activities that poses major threat for its longevity and in turn to the survival of the lotus and other biological resources (Jain *et al.*, 2004).

The lotus plant is perennial and grows naturally in subsequent years so the farmers do not give much attention to its cultivation. In Manipur lotus cultivation is generally through the rhizomes mainly during February and March. Cultivation through natural seeds is also in practice. Large numbers of seedlings emerge from seeds already buried inside the mud, which can be collected as fresh planting material.

HABITAT AND ECOLOGICAL ENVIRONMENT

CHAPTER-9

Habitat

The lotus (*Nelumbo nucifera* Gaertn.) is a wetland macrophyte species. It is one of the most important aquatic plant species in the Indian sub-continent. It occurs widely in subtropical, tropical, sub-temperate and temperate regions throughout the country in fresh water. The lotus generally abounds in fresh water reservoirs, shallow lakes and small ponds. Throughout the Indian sub-continent, there are many shallow water bodies with an average depth of less than two metres where this species flourishes. Growth of the floating leaved plants is governed by the ability of their petioles to elongate as per the water depth. *Nelumbo* is well known for the elongation of its petioles by intercalary meristematic growth following a rise in the level of water. Aquatic weeds, such as *Eichornia crassipes* (Mart.) Solms 'Water Hyacinth', drastically check the growth of this species. The white flowered *Nelumbo* in particular is eradicated by the invasion of such weeds.

Lotus (*Nelumbo nucifera*) in full bloom

In the Odisha region, the multipetalled white *Nelumbo* is quite common, whereas in West Bengal, multipetalled pink and white *Nelumbo* races have been located mainly in the Midnapur region. Clay soil rich in humus and shallow water forms the ideal habitat for this species. It is interesting to note that it is able to grow in a wide range of pH (7.5–9.0) but during its growth brings the pH (7) to neutral. Eutrophication by agricultural runoff and domestic sewage is commonly held responsible for the explosive growth of this species. During course of exploration, *Nelumbo nucifera* with pink flowers was recorded from a saline lake (pH 9.3) near Samaspur in the Raebareli district of Uttar Pradesh, showing the variable range of habitats adopted by this species. However, experimental

Lotus in saline lake (pH 9.3) at Samaspur, Raebareli (UP)

data on the effects of different levels of salinity on growth and the tolerance limits are not available. It has also been observed during investigations that it possesses the ability to take up heavy metals so it can be grown in the ponds receiving industrial effluents, and thus serving the purpose of water purification in aquatic bodies.

This species is of great ecological amplitude and occupies a wide range of habitats. It adapts well to the different types of environment by the phenotypic plasticity (morphological and/or physiological) showing polymorphic variations. It is found in a wide range of geographical distribution in different climatic zones, and it is quite likely that different ecologically distinct populations in terms of races represent this species with large variations in flower colour, number of petals and other significant morphological characteristics.

The wetlands and aquatic bodies comprise one of the most important habitats for the preservation of balanced ecosystems in this country. The Indian subcontinent harbours nearly 14,50,800 hectares of natural wetlands. During the last five decades, aquatic habitats are deteriorating rapidly due to demographic pressure, urbanization, invasion of aquatic weeds and increased inflow of industrial effluents causing environmental degradation to an alarming extent (Mohan Ram, 1991). Due to environmental degradation, the aquatic

Flowering of Lotus in Nawabganj Lake, Unnao (UP)

Flowering of Lotus (pink) in a pond near Sitapur (UP)

bodies are shrinking rapidly causing severe genetic erosion in the diversity of this taxon. As a result, *Nelumbo nucifera* has been under tremendous pressure for the reasons, and requires immediate attention on its conservation.

Ecological Environment

Nelumbo is a typical wetland plant species and water is indispensable for its natural ecological environment. The lotus is a long-day plant and likes moderate conditions of light, humidity and temperature during the growth season. July is the best time for the flowering of the lotus. The lotus grows from Kashmir to Kanyakumari which indicates that it is a highly versatile plant. Deep water is not suitable for the growth of lotus plants. The plant can usually tolerate water up to 1.5 m in depth. If the water level is more, the standing leaves will emerge and the plant will die after sometime. The seeds disposed in the area can protect themselves from the impact of adverse conditions and once the conditions are favourable, seeds start germinating to keep the continuity and survival of this species. The dispersal of lotus seeds is carried out by water in floods.

Water: The lotus prefers calm and shallow water such as a pond and lake. The ideal depth of the water level is 1.0 m. When the leaves dry up and dormancy starts, the lotus needs very little water but not the complete dryness.

Soil: Light clay with rich organic matter is ideal for lotus cultivation. Soil with 6.5-7.5 pH is preferred.

Temperature: For the production of lotus flowers, the ideal temperature is 22-35^{0}C.

Light: Lotus is a long day plant and needs atleast eight hours sunlight. Lotus and fish live comfortably together and maintain a good ecological balance and Productivity.

MORPHOLOGICAL AND ANATOMICAL CHARACTERS OF LOTUS

CHAPTER-10

Rhizome

The submerged stem of *Nelumbo* spreads horizontally. This rhizomatous stem stores food and helps in the asexual reproduction. The rhizome of the lotus is interesting on account of its vascular anatomy. The structure in the cross section shows epidermis, hypodermis, parenchyma and vascular bundles. There is no stele and a large number of vascular bundles are scattered in the parenchyma. Vessels are absent in the lotus rhizomes and the vascular bundles have no cambium. Fibre is present in the tracheidal wall of the vascular bundles which is very elastic.

The first growth phase is vegetative in which each leaf petiole that emerges from a node is larger than the previous one. When the new petioles are shorter than the older ones, the plant switches to producing rhizomes (Yamaguchi, 1990). Rhizomes grow nearly 10 m in a year. During the early growth of the rhizome, the rib meristems contribute mainly to the internode. As a result of these divisions the internode becomes elongated. The tuberization of the rhizome is built up from cell divisions of three kinds of tissues:

- ✓ Primary thickening of meristems
- ✓ Cells of the vascular bundles
- ✓ Parenchyma of cortex

But the growth in thickness of the rhizome seems to be chiefly due to the enlargement of the parenchymatous cells (Chen & Zhang, 1989).

Lotus rhizomes which grow several meters long

Lotus Root

In the seed-grown plants the main root is developed from the radicle. On the rhizome node, adventitious roots come out which anchor the plants. Roots absorb water and nutrients for the growth of the plants.

Lotus Leaf

Leaves are peltate, orbicular, entire and glaucous 30-60 cm in diameter. Petioles are long, hard and thick with prickles. The leaf is the photosynthetic organ which manufactures food for the plant. The large green circular leaves are unique in their water resistance and self-cleaning properties. Water droplets become rounded on the surface of the leaves and demonstrate mercury-like characteristics as they roll off the plant. Lotus plants by contrast have their stomata on the upper surface of the leaves, exposed to the air. The petiole is cylindrical 20-250 cm long and 0.5-1.5 cm in diameter with prickles. Leaves are grouped into two forms. First the floating leaf because it cannot stand up due to the soft petiole. The second one is the standing leaf which comes out with a tall, thick and hard petiole.

Standing leaf with hard petiole in Lotus growing in pot

Lotus flower with calyx, corolla and fruit (torus)

Lotus Flower

In most of the lotus cultivars, the flower bud begins to differentiate when three standing leaves come out. It takes 4-6 days for the flower bud to complete differentiation. Once the flower bud appears above the water, it indicates the end of differentiation. The flower is solitary and hermaphrodite. It comprises calyx, corolla androecium and gynoecium. The flower bud is long, peach-shaped and varies from 10 to 30 cm in size. It is green, pale yellow, pink and red as per the flower colour of the cultivar. The outermost whorl of the flower is the calyx with 6-7 sepals. They are thick, tenacious and green. The corolla is composed of many petals varying enormously from 8 to 1000 in number according to the cultivar. Petals are almond shaped and they are the most ornamental and elegant part of the flower in white, pink, red and pale yellow shades as per the cultivar. The showy petals attract the insects, especially honeybees which help in the cross pollination. The flowers of *N. nucifera* races are more attractive than the pale yellow flowers of *N. lutea*.

***Nelumbo lutea*: Yellow Lotus**

***Nelumbo nucifera*: Pink Lotus**

Lotus Fruit and Seed

The lotus fruit is the nut. The fruit is green when immature and turns brown at maturity. Seeds are hard and dark brown. They vary in shape from round to oval to oblong. Lotus seeds are the oldest viable seeds ever recorded among the angiosperms. Seeds can lie dormant for centuries and then grow into a plant in favourable conditions.

Seeds are 1.2-2.0 cm in length and 1.1-1.5 cm in diameter. The average weight of a seed is 1 gm. The Pericarp of a lotus seed is developed from the ovary wall only. It is differentiated into exocarp, mesocarp and endocarp. Inside the lotus seed we can find edible endosperm. The first leaf is formed like the neck of a swan in dark green colour. It is an unusual phenomenon that, in the absence of light, the lotus embryo develops such a dark green in colour.

Lotus - Oldest Viable Seed

Recent research using carbon-dating techniques have shown that the lotus seed is the oldest demonstratively viable seed ever reported

Dark green embryo and seeds in Lotus

among the angiosperms. Scientists are now using evidence from lotus seeds to understand the ageing process of other species. A lotus seed can withstand thousands of years without water. A lotus seed can withstand thousands of years without water.

In the 1920s, some lotus seeds were discovered from the lake sediments in Northeast China. No one knew how old the seeds

Morphological variation in petals' size

were at that time. They were planted and successfully germinated into healthy plants. But it was only in the 1990s that the age of these germinated seeds was revealed. Astonishingly, they were found to be around 1300 years old. Scientists believe that this extraordinary ability to remain viable for such a long period is due to a special enzyme called L-isoaspartyl methyltransferase. The enzyme is responsible for the seeds repairing damage to proteins within their cells before germination. This could contribute to our knowledge of the aging process in the other organisms, and scientists are following up research in this field (Shen-Miller, *et al.,* 1995; Shen-Miller, *et al.,* 2002).

The present investigation relates to the use of the sacred lotus seed extract in cosmetic, dermatological and/or dietary compositions for combating aging and methods for the treatment and prevention of aging by using such a composition. It has been observed that using an extract of sacred lotus seed by topical application as an active agent, signs of aging can be reduced, eliminated or even reversed. Lotus seed has the ability to reduce the appearance of wrinkles as well as nourish the body with powerful age-defying antioxidants. Studies are in progress to develop and standardize the nutraceuticals by using seed kernels in Indian and Chinese laboratories (Wu, *et al.,* 2007). Nutritional composition, physiological functions and processing of lotus seeds have been reviewed in detail by Zhang, *et al.*, (2015).

Seeds used to date others

Lotus with matured stamens and torus ready for fertilization

TAXONOMY AND NOMENCLATURE OF LOTUS

CHAPTER-11

Nelum (නෙළුම්) *is* the Sinhalese name for the Sacred Lotus. In Latin *Nelumbo* means "Sacred Beans" and *nucifera* means "Having hard fruit". During the course of systematic studies by various workers, the genus *Nelumbo* has repeatedly been attributed different taxonomic positions. Dumortier (1829) and Lindley (1836) elevated *Nelumbo*, (previously included in Nymphaeaceae) to the rank of a separate family Nelumbonaceae. Arber (1920) stated that *Nelumbo* should be interpreted as a highly differentiated genus among the water lilies. Most probably this genus has descended from aquatic ancestors while it reverted to some degree towards terrestrial life. Li (1955) considered *Nelumbo* to be so aberrant that he proposed a separate order Nelumbonales for this genus. Takhtajan (1959) supported the view of Li (1955) and mentioned the tricolpate pollen grains, the morphology of the gynoecium, the absence of suspensor and of perisperm, the massive embryo with semi-globose cotyledons, the presence of a special respiration pore, the germ plumule, the lack of ideoblasts and the different chromosome number (2n=16). Khanna (1965) while studying the morphology and embryology of Brasenia and *Nelumbo nucifera* made an attempt to emphasize the differences between these two taxa, such as the horseshoe shaped carpel of *Brasenia* and the barrel shaped carpel of *Nelumbo nucifera* and cellular endosperm in the former and nuclear in the latter. On the basis of these characteristics she justified the view of Li (1955) and Takhtajan (1959), that *Nelumbo* should be placed under a separate family and raised to the status of order Nelumbonales.

A double white flower of *Nelumbo*

Ito (1987) applied a cladiastic analysis to reveal the phylogenetic relationship among the Nymphaeales. He concluded that *Nelumbo* is a distinct taxon and is presumed to have originated from an ancestral stock of Nymphaeales. Therefore, genus *Nelumbo* of the family Nelumbonaceae should be placed under the order Nymphaeales. Further, he has also discarded thesystem of Thorne (1983), in which he grouped family Ceratophyllaceae with Nelumbonaceae to the order Nelumbonales. According to Ito (1987) the segregation of *Ceratophyllum* from the Ceratophyllum group (*Cabomba, Brasenia and Ceratophyllum*) would result in a paraphylatic situation of the subfamily Cabomboideae of the Nymphaeaceae and the system dealing with such paraphylatic relationships between the families/subfamilies is not acceptable (Angiosperm Phylogeny Group, 2016).

Genus *Nelumbo* comprises two species viz., *Nelumbo nucifera* Gaertn. and *Nelumbo lutea* (Willd.) Pers. Besides these two species, a number of cultivars and races of both these species occur widely throughout the world in different geographical regions.

The name *Nelumbo* can be found for the first time in Tournefort's Institiones Rei Herbariae, which according to Stafleu & Cowan (1976-88), was first published in 1700. As a generic name *Nelumbo* was then validly published in 1763 by Adanson in his "Familles des Plantes". Consequently the type of *Nelumbo nucifera* Gaertner, which is identical with the type of genus originally named *Nymphaea nelumbo* Linnaeus. In 1788 Gaertner added in his monograph of fruits and seeds the specific epithet "nucifera" in the sense of the Linnaeus binary system. He validly published a new name for the plant *Nelumbo nucifera* Gaertner.

Willdenow published *Nymphaea lutea* in 1798, which is an unambiguous name for the new world taxon on the species level.

Nelumbo (Lotus) is the only genus in the family Nelumbonaceae superficially similar but unrelated to the water lilies of the family Nymphaeaceae. *Nelumbo* is what is commonly meant by the rather ambiguous term, Lotus, though as a scientific name. *Lotus* refers to a completely different unrelated genus in the Legume family.

The leaves of *Nelumbo* can be distinguished from those of the genera under Nymphaeaceae as they are peltate and fully circular. *Nymphaea* on the other hand, has a single characteristic notch from the edge to the centre of the leaf. The seed pod is conical like a water rose is also a distinct character of *Nelumbo*.

According to the common view the genus *Nelumbo* comprises two extant species: the Asian *Nelumbo nucifera* Gaertner having pink, reddish to white flowers and the North American *Nelumbo lutea* (Willd.) Persoon pale yellow flowers. With morphological, anatomical, palynological and phytogeographical investigations carried out on extensive material from the whole range of the genus on living and herbarium specimens, it is shown that both taxa are hardly differentiated. After discussing the nomenclature and historical development of systematic concepts in *Nelumbo* the New World taxon is now considered to be a subspecies (*N. nucifera* subsp. *lutea* comb. et stat. nov.). (Borsch & Barthlott, 1994.)

The occurrence of tubular wax crystals containing a significant amount of noncosan -10-0l supports a position near Ranunculiflorae. A unique new feature of *Nelumbo* is the occurrence of aporphin alkaloids (nuciferin, nornuciferin) in cuticular waxes. These alkaloids, well known from the tissues of *Nelumbo*, are common in Ranunculiflorae, whereas Nymphaeales are characterised by sesquiterpene alkaloids (Barthlott, *et al.*, 1996).

Nelumbonaceae

This is a monogeneric family with only two species, *Nelumbo nucifera* Gaertn. and *N. lutea* Pers. Herbs, perennial, aquatic, rhizomes branched, form swollen terminal storage rhizomes late in the growing season. Leaves arising from rhizomes alternate, immersed or floating, long petiolate, leaf blade centrally peltate, veins radially extended. Flowers are solitary, axillary, long pedunculate, bisexual, hypogynous, actinomorphic, entomophilous, borne above the water surface. Tepals are numerous, distinct, outermost reduced, inner ones larger and petaloid. Stamens are numerous, anthers dehiscent by longitudinal slits. Pistils are numerous, simple, separately and loosely embedded in cavities on the flattened top of the receptacle. The ovary is 1-loculed, ovule 1, pendulous, styles very short. Stigmas are capitate. Fruit are nut like, indehiscent. Seeds are without endosperm and perisperm, embryo large, cotyledons - 2, fleshy

Enumeration

Nelumbo Adans.

Adans., Fam. 2: 76. 1763; Wood in J. Arnold Arb. 40: 105. 1959. *Nelumbium* Juss., Gen. Pl. 68. 1789 'orth var.'; Benth. in Benth. & Hook. f., Gen. P1. 1: 47. 1862; Hook. f. & Thoms., F1. Ind. 1: 248. 1855 & in Hook. f., F1. Brit. India, 1: 115-116. 1872; Caspary in Engl. & Prantl, Nat. Pflanzenfam. 3, 2: 5. 1888; Mitra, Fas. F1. India, 20: 8. 1990.

Perennial, large, rhizomatous, laticiferous, aquatic herbs have scattered vascular bundles with air-canals. Rhizomes are branched, stoloniferous, creeping, tuberous or slender with adventitious roots at nodes, solitary leaf, flower and axillary buds. Leaves are long petiolate, simple, peltate, stipulate, orbicular with radiating veins branched dichotomously except the midrib; juvenile leaves eccentrically peltate, floating; matured ones centrally peltate, floating or immersed. Flowers solitary on long peduncles, raised above the water surface, large showy, white to rose-pink, red or yellow, bisexual, hypogynous with spirally arranged floral appendages. Sepals 4-5 in number, free, subequal, petaloid, persistent. Petals are numerous, free, intermediate ones largest, caducous; inner ones sometimes transitional to stamens. Stamens are numerous, free, long, linear, differentiated into subterete filaments; anthers with adaxial to sublateral microsporangia have a distal, clavate, incurved sterile appendage; pollens tricolpate. Carpels (9-) 12-28 (-40), distinct, are embedded singularly in cavities on turbinate, accrescent receptacles. Ovules are solitary in each carpel, pendulous, orthotropous, becoming anatropous after attaining maturity, bitegmic, crassinucellate with laminar placentation. Nuts ripen above the surface of the water; pericarp smooth and bony. Seeds are in the carpel, exarillate, inoperculate; cotyledons large, fleshy; plumule folded, enclosed in a stipulate sheath.

Nelumbo nucifera Gaertn.

Nelumbo nucifera Gaertn., Fruct. 1: 73 t. 19. f. 2. 1788; Biswas & Calder, Handb. Wat. Mar. P1. Ind. Burma, ed. 2, 23. 1955; Subramanyam, Aquat. Ang. 8. f. 4. 1962; Santapau in RBSI, 16 (1): 6. 1967; Mitra, Fas. F1. India, 20: 8 - 10. 1990. *Nymphaea nelumbo* L., Sp. P1. 511. 1753. *Nelumbo indica* Poir. in Lam., Encycl. 4: 543. 1798. *Nelumbium speciosum* Willd., Sp. P1. 2: 1258. 1799; Wight, I11. 1: t. 9. 1840; Hook. f. & Thoms., F1. Ind. 1: 247. 1855 & in Hook. f., F1. Brit. India, 1: 116. 1872; Roxb., F1. Ind. 1874 (Repr. ed. 450-451, 1932); Sims in Curtis, Bot. Mag. t. 903. 1806; Hook. f. in Curtis, Bot. Mag. 68: tt. 3916 & 3917. 1842; Cooke, F1. Bombay, 1: 26. 1901 (Repr. ed. 1: 28. 1958); Prain, Beng. P1. 1: 214. 1903 (Repr. ed. 1: 141. 1963); Duthie, F1. Upp. Gang. P1. 1: 35. 1903; Gamble, F1. Madras, 1: 34. 1915 (Repr. ed. 1; 24. 1957); Haines, Bot. Bihar & Orissa, 1: 22. 1921 (Repr. ed. 1: 23. 1961); Kanjilal *et al.*, F1. Assam, 1: 65. 1934. *N. nelumbo* (L.) Druce in Rep. Bot. Exch. C1 3: 421. 1924.

Nelumbo nucifera Gaertn. (Sacred Lotus)

An aquatic perennial; with leaves suborbicular, 20-80 (-100) cm across, they lie flat when floating and cupped when emerged, glabrous and glaucouson both surfaces; petioles 1-2 m long fistulous beset with hard minute papillae-like structures; flowers 8-25 (-30.0) cm across are rose-pink to white and fragrant; sepals 1.5-5.5 x 0.8-3.5 cm are ovate to elliptic, concave, green (in white flowers) or pinkish-green (in rose-pink flowers). Petals are 16-24 (in single forms) or 116-160 (-198) in double forms with stamens up to 230, 2.2-5.0 cm long; outermost staminodial in double forms and receptacle 5.0-10.0 cm across, spongy, yellow during anthesis; stigma papillose. Fruit is oblong to ovoid, glabrous, pericarp thick, hardened, 1.0-2.0 cm in length and 1.0-1.5 cm in diameter. Flowering occurs from in May to October (Sharma & Goel, 1997).

Nelumbo lutea (Willd.) Persoon ('American Lotus')

The American lotus, *Nelumbo lutea* (Willd.) Pers., is native to Colombia, North America. It is widespread in Eastern and Central North America and also found in Central Canada. It is also known as the New World Lotus. It has thick spreading rhizomes and large bluish green circular leaves which are usually raised above the surface of the water. The pale-yellow flowers usually open for three successive days. Both the seeds and rhizomes are eaten by the Native Americans. The American lotus is considered sacred by the Native American cultures and is said to have mystic powers (Ward, 1977).

Nelumbo lutea Pers. ('American lotus', 'Water Chinquapin', 'Yanquapin')

This is an aquatic perennial; rhizomes are stout and banana shaped and can reach up to 25 cm in length; leaves peltate, usually raised out of the water, cup shaped or depressed at the centre over the point of attachment of the petiole, 0.30-0.90 m in diameter, bluish green above, light yellowish on the lower surface at the young stage; petioles 1.5-2.0 m long, papillate; flowers 12-25 cm across on 1.5-2.0 m long pedicels, pale sulphur yellow like a double tulip; petals ovate-obtuse; anthers hook appendaged; seeds acorn-like, 1.5 x 1.30 cm (Bailey & Bailey, 1976; Everett, 1980).

Nelumbo lutea Pers. was introduced in India (NBRI, Lucknow) for the first time from the Missouri Botanical Garden, USA (Sharma & Goel, 1997).

A white lotus bud

HORTO-TAXONOMY OF LOTUS VARIETIES

CHAPTER-12

All cultivated plants have arisen from their wild relatives. During the course of their introduction, domestication, rampant and planned hybridization and mutation breeding, a large number of cultivars have been evolved and released in the horticultural trade. There is considerable diversity in their pattern. The importance of Horto-taxonomy (Taxonomy of the cultivated plants) lies in the fact that it helps to arrange properly all the variations of cultivars in a most meaningful manner for the purpose of their correct identification and nomenclature. For the utilitarian purpose, taxonomy based on morphological differentiation is more relevant and useful, even though taxonomy based on genetic relationship may be more phylogenetic (Khoshoo, 1971 & 1980).

According to the International Code of Nomenclature for Cultivated Plants (Brickell *et al.,* 1980), a cultivar is the basic unit and no subdivision of this category is possible. A cultivar should not be confused with a botanical variety. Cultivar has been defined under Article-10 of the code as 'an assemblage of cultivated plants which is clearly distinguished by any character (morphological, physiological, cytological, phytochemical or others) and when reproduced (sexually or asexually), retains its distinguishing characters'. Cultivated plants generally do differ in many respects from their typical wild plant species. The formal botanical categories can successfully accommodate both wild and cultivated plants down to the species level but they are inapplicable to the intraspecific levels for the systematics of cultivated plants (Jeffrey, 1968).

Concept of Cultivar

The concept of a cultivar is essentially different from that of the botanical variety. The latter is a category below the species level. Names of botanical varieties are always in Latin and are strictly governed by the International Code of Botanical Nomenclature (ICBN) (Gruter *et al.,* 1988). The current concept of the category 'cultivar' is recognized in origin, history and biological significance, which is markedly distinct from a botanical variety. The exact meaning of the latter concept differs somewhat from botanist to botanist, but usually designates a natural interbreeding, more or less self perpetuating population. The cultivar, on the other hand, owes its distinction to one or more characters that are usually maintained in the population artificially, by asexual propagation (where the plant is a clone), by controlled breeding or by continued selection to an acceptable standard (Lawrence, 1963).

Horto-Taxonomy

The methods of formal taxonomy have not been satisfactory for classifying cultivated plants. As a result, people who deal with cultivated plants, most of them have developed their own informal and intuitive classifications based on their experience of useful groupings. The structure of the total gene pool is characterized by assigning taxa to primary, secondary and tertiary gene pools. At the intraspecific level, cultivars are grouped into races and subraces in an informal way without rigid rules for the use of terms (Harlan & De Wet, 1971).

A classification involving cultivated plants is not to stress upon botanical differences, but to take into account their horticultural use and resolve cultivars into groups to fulfill the purpose in the field, garden or a need in an exhibition. Most of the classifications are, therefore, based on characters of the particular parts of the plant developed for humans' utilization as criteria for the separation of cultivars. It has been realized that there has to be consistency in the classification of cultivated plants to enable easy identification of cultivars. This would lead to correct documentation of germplasm collection for the purpose of conservation and use in the plant breeding programmes (Khoshoo 1971 & 1980). Thus, Horto-taxonomy can be defined as studies on identification, classification, nomenclature and usage of cultivated plants (cultivars) in a meaningful manner.

The first International Symposium on the Taxonomy of Cultivated Plants at Wageningen in August, 1985, drew a large number of botanists and horticulturists who contributed a great deal to the existing knowledge and learnt from the detailed discussions. Each cultivar needs to be registered and typified with an authentic voucher specimen, description and coloured photograph, year and place of origin, introduction and mention about its ancestry.

The problem is more complicated at the level of cultivars, as rules of nomenclature are not strictly followed. Lack of correct pedigree records and other relevant data before the cultivar is designated or named has not only created confusion but given rise to an enormously large number of cultivars of uncertain origin/identity not meriting their separate placement (Arora *et al.*, 1980).

Conspicuous variation within taxa generally prevails, often crossing their taxonomic limits, thereby posing problems for distinct categories. Thus, cultivated plants are different from wild ones, meriting special consideration and suitable guidelines for appropriate grouping. Fixity of names to the cultivars and designating them as valid taxa as per rules of the International Code is important so that stability may prevail. Even when, at a later stage, a particular cultivar becomes obsolete, it can be identified by its earlier validly published authenticated record. A critical study of comparative morphology is a very helpful tool in designating the status of a taxon of ornamental significance.

There is tremendous scope and demand for the authentically identified material of ornamental cultivars/varieties of plants in the horticultural trade at the present time. Detailed horto-taxonomic studies of ornamentals will serve as an important tool in order to check the corruption of unnecessary naming of cultivars just for the sake of name and fame and for boosting the sale of plants by nurserymen. The solution lies in intertwining botany and horticulture for authentic identification, appropriate nomenclature, classification and various uses of the cultivars practised in the horticultural trade in the present day patent regime (Sharma & Goel, 2006).

Horto-Taxonomy of *Nelumbo* Cultivars

Taxonomists, concentrating on the systematics of ornamental plants, accept the requirement that to achieve widespread recognition and acceptance, they must translate the findings of horto-taxonomical investigations into terms understandable and useful to nurserymen and plant lovers. Similarly, the authentic germ plasm collection of *Nelumbo* cultivars needs to be screened and recommended for various purposes in horticulture and landscaping according to habit, growth and flowering behaviour.

Horto-Taxonomic Characters of *Nelumbo* Cultivars

The following parameters for the horto-taxonomical studies on *Nelumbo* cultivars may be considered for the preparation of their descriptions.

Rhizome	:	Colour and size
Leaf	:	Colour and size
Petiole	:	Colour, size and density of prickles
Bud	:	Colour and size
Flower	:	Colour and size
Sepal	:	Colour, shape, size and number
Petal	:	Colour, shape, size and number
Stamen	:	Number and size
Filaments	:	Colour and size
Anther lobes	:	Colour, shape and size
Carpellary	:	Colour and size of receptacle
Carpel	:	Colour, shape, size and number
Seed	:	Colour, shape, size and number

For the coloured documentation of races, the Horticultural colour charts I & II p. 200, published by the British Colour Council in collaboration with the Royal Horticultural Society, London (Wilson, 1938) may be consulted.

Nelumbo nucifera 'Krishna'

A unique and charming 160 petalled pink lotus race from Midnapore, West Bengal was discovered and introduced into the Botanic Garden of the CSIR-National Botanical Research Institute, Lucknow, in 1996. This race has been named *Nelumbo nucifera* 'Krishna' (Kamal Krishna). The cultivar has been acclimatized, multiplied and widely disseminated. Salient features are described below:

Rhizomatous aquatic herb; Leaf 40-55 cm diameter, veins 21-22 in number; Buds 10.5-12.5 cm long; Flowers, rose madder, 22-26 cm diameter; Sepals four in number, lettuce green; Petals 116-160 in number, multiseriate, rose madder, inner petals staminoid, 3.1-9.0 cm long; Stamens 220-250 in number; Carpels mimosa yellow 14-21, ovule solitary; Seed 1.4-1.7 cm long, dark grey. Flowering and fruiting from April to October; Propagation by rhizome and seed (Sharma, *et al.,* 1998).

Propagation by rhizome in Lotus

Germinating seeds of Lotus

Lotus Cultivars for Various Uses

Size, shape, number and colour vary from cultivar to cultivar. Large flowered lotus cultivars are up to 30 cm in diameter and the small-flowered cultivars are less than 10 cm in diameter. Number of petals range from 8 to 1000. Petals are white, pink, rosy red, red, light yellow and bi-coloured. The shape of the petals varies from elliptic to long elliptic and are arranged spirally at the base of the receptacle. The single lotus flower transforms into double forms from stamens to stamenoids and finally to the petals.

Studies have shown different lotus cultivars perform differently in their growth habits. Some cultivars produce good flowers but no seed or rhizomes whereas some produce good rhizomes but without any or with very few flowers and seeds. For instance, 'Quandong' is the only cultivar that is suitable for rhizome production. Cultivars 'Singapore' and 'Brisbane' produce good rhizomes. 'Green Jade', 'Vietnam-Red' and 'Brisbane' are excellent cultivars for the flower production. Lotus cultivars have been classified into three categories according to their usage: rhizome, flowers and seeds. Some cultivars may exhibit one or more of the three characters but generally each is classified by its salient feature. There are nearly 400 *Nelumbo* cultivars all over the world. Lotus flowers are termed single when they have less than 25 petals, semi-double if they have 25 to 50 petals and double if they have more. Lotus cultivars have also been classified by height: small, medium and large. The other way to classify cultivars is by their colour: red, pink, white, pale yellow and their various shades. Lin *et al.* (2019) has conducted detailed studies on *Nelumbo nucifera* cultivars and has concluded that this taxon has potential to be an emerging horticultural model plant. It also has lot of economic and medicinal potential.

***Nelumbo nucifera* 'Chavan Basu' in an aquatic body**

***Nelumbo nucifera* c.v. "Chavan-Basu"**

FLORAL BIOLOGY OF THE LOTUS

CHAPTER-13

It is observed that the floral structure (i.e. large numbers of stamens, masses of pollen, staminal appendages) are adaptations which facilitate pollination of *Nelumbo* by beetles. The most common insect visitors to the lotus flower are honeybees (*Apis melifera*). Anthesis occurs over a period of three days. First day flowers are protogynous. Anthesis of first day flowers is indehiscent and inaccessible to insects, since they are pressed tightly against the sides of the receptacle by the perianth parts. There is a difference of time between stigma receptivity and pollen release from the anther within the same flower which avoids self-pollination. Flowers open in the morning and at first only female parts are exposed. This helps ensuring cross-pollination because flowers are likely to be pollinated from older flowers where pollen is already being released. Seed formation is most efficient when pollen from the first day old flowers is crossed with pistils from first day flowers (Kasumi & Sakuma, 1998).

Overall floral structure coupled with the sequence of floral development, odour and timing of perianth movements (opening and closing) are adaptations which facilitate the pollination of *Nelumbo* by beetles (Schneider & Buchanan, 1980). Cross pollination helps in widening the genetic diversity and strong offsprings. The lotus has an entomophilous flower and requires insects to gather pollen for its fruits to ripen. It is not correct to say that in the lotus flower and fruit formation happen at the same time.

In most lotus cultivars, flowering lasts for four days. On the first day the flower starts opening, second and third day keep blooming and on the fourth day starts withering. Lotus flowers take 7-8 hours from pollination to fertilization. After 10-15 days of fertilization, the receptacle expands. Depending on the weather, the lotus takes 30-35 days from pollination to developing fruits and seeds. The fruits are soft and green which gradually change to light brown and become hard on maturity. After ten days fruits mature well with brown and hard skin. Gradually the fruits separate from the receptacle which is the best time for harvesting them. If delayed, the nuts will fall from the receptacle into the water for dispersal.

Matured stamens and carpels

Pollination to Seed Formation (30-35 Days)

First day	:	Flower opens
Second day	:	Pollination and fertilization (7-8 hours)
Second and third day	:	Blooming
Fourth day	:	Withering
10-15 days	:	Receptacle expands and full of green fruits
30-35 days	:	Fruits and seeds get matured

Lotus is a long day plant and prefers warm conditions. The optimum temperature for growing lotus flowers is 22-35^{0}C, which influences the growth and development of the lotus flower. Study shows that temperatures lower than 18-20^{0}C are harmful to fertilization and thus affect fruiting as well. It develops faster in sunny locations than in shady positions. Lotus grows well in 20-30 cm thick light clay soil with rich organic matter in shallow water.

A young torus after fertilization

***Nelumbo nucifera* (close-up of matured pistils)**

LOTUS EFFECT

CHAPTER-14

For centuries, Indian religions have recognized the purity of the lotus leaf. This purity is caused by its super-hydrophobic surface, which allows water drops to form spherical balls that collect the dust particles as they roll off with very little contact resistance.

All primary parts of plants except roots are covered by a cuticle that is the interface between plants and their environment. The cuticle is composed of soluble lipids embedded in a polyester material (Holloway, 1994). Due to its chemical composition, the cuticle in most cases forms a hydrophobic surface.

The micro-relief of plant surfaces, mainly caused by epicuticular wax crystalloids, serves different purposes and often causes water repellency. Furthermore, the adhesion of contaminating particles is reduced. Interdependence between surface roughness, reduced particle adhesion and water repellency is the key factor in the self-cleaning mechanism of many biological surfaces. The leaves of the lotus (*Nelumbo nucifera* Gaertn.) afford an impressive demonstration of this effect which is called "The Lotus Effect". This effect is of great biological and technological importance (Barthlot & Neinhuis, 1997).

Lotus Effect on a leaf

Lotus Effect on Pink double lotus leaves

Scanning electron microscopic studies of biological surfaces have revealed an incredible micro structural diversity of the outer surfaces of plants. Microstructures, such as trichomes, cuticular folds, and wax crystalloids, serve different purposes and often provide a water repellent surface. Water repellency is mainly caused by epicuticular wax crystalloids which cover the cuticular surface in a regular micro-relief of about 1-5 µm in height (Jeffrey, 1986).

From a number of experiments, it was proved that water repellency causes an almost complete surface purification (self-cleaning effect). Contaminating particles are picked up by water droplets or they adhere to the surface of the droplets and are then removed with the droplets as they roll off the leaves. Water repellent, rough surface properties were represented by the leaves of the lotus (*Nelumbo nucifera*). Depending on the wettability, applied water droplets displayed distinct characteristics when rolling off the leaves. On water repellent surfaces, water contracted to form spherical droplets. In general, particles of any kind were removed entirely from water repellent leaves when subjected to natural or artificial rain, independent of their size and chemical nature, as long as the surface waxes were not destroyed. This can be seen most impressively with the large peltate leaves of the lotus. According to tradition in Asian religions, the sacred lotus is a symbol for purity, ensured from the same observations made scientifically. This knowledge is well documented in Sanskrit writings and the phenomenon is now known as "The Lotus Effect".

Due to the very small interfacial area between particle and rough surface, adhesion is minimized. Therefore the particle is captured by the water droplet and removed from the leaf surface. The effectiveness of the self-cleaning ability decreased in fog or dew, in contrast to rain. Rain droplets have a high kinetic energy. Their elastic formation allows them to penetrate between epidermal papillae and remove the particles within the troughs. This does not happen with fog and especially not with dew.

Biological Implications of the Lotus Effect

The cuticle is the outermost barrier of plants towards their environment and is, therefore, the first protective layer. Because the air contains many kinds of particles, leaf surfaces are continuously contaminated. It was shown that in polluted areas where plants are heavily contaminated with dust, the leaf surface temperature increased under isolation (Eller, 1977). Particles within a certain size range occlude stomata and influence stomatal diffusive resistance. Water repellent plants escape from those harmful effects through the Lotus Effect. The Lotus Effect plays another important role in the defence against pathogens. Spores and conidia are deprived of the water which is necessary for their germination (Campbell *et al.*, 1980; Juniper, 1991). Therefore, the epicuticular wax crystalloids and their physical properties may be regarded as the first line of defence against pathogens.

The Lotus Effect is not restricted to plants; indeed, it is of overall biological significance, e.g. for insects with large wings, which cannot be cleaned by legs, have a water repellent property, using surfaces which exhibit the self cleaning ability (Wagner *et al.*, 1996). In this case, not only the removal of particles is of interest, but also the maintenance of flight capability, which may be lost due to an unequal load on the wings.

The self-cleaning mechanism may be the most important function of many micro structured biological surfaces. This effect can be transferred to artificial surfaces (e.g. cars, facades, roof tops and walls) and innumerable other biological and technical applications.

Nowadays, water-repellency has gained keen interest because it represents the basis for the self-cleansing property of such surfaces called the "Lotus Effect" (Barthlot & Neinhuis, 1997). The German scientists have developed paint, "Lotusan", which mimics the lotus leaf and repels dust particles and water droplets keeping the buildings clean (Marmur, 2004).

The special feature of lotus leaves allows water and dirt particles to run off from the leaf surface. The Lotus Effect is due to characteristic molecular structures present on the surface of the lotus leaf which ensure that water drops and dirt remain on the tips of the plant structures, minimizing the leaf contact area with foreign objects. This property is being applied to developing self-cleaning textiles. A project has been undertaken in co-operation with NEES-Institute for Biodiversity of plants at the University of Bonn and BASF. Such textiles would need only water for cleaning, thus saving money and time on washing. Textiles utilizing the Lotus Effect would also be more resistant to wear and tear.

Another application of the Lotus Effect is in medicine; scientists at ITV (Institute for Textile Technology and Process Engineering) are investigating how they could alter the surface of implants in order to improve the implants coating with human cells.

Nelumbo nucifera cv.'Krishna' pink double lotus' in an aquatic body exhibiting the Lotus effect

Close-up of _Nelumbo nucifera_ (Pink) in aquatic body exhibiting the Lotus effect

A white pinkish lotus bloom with pink stripes on petals

THERMO-REGULATION OF LOTUS FLOWER

CHAPTER-15

The lotus flower works like a thermostat. The 'sacred lotus', *Nelumbo nucifera* Gaertn., not only produces a significant amount of heat during the sequence of flowering but also regulates its temperature precisely. If the air temperature varies between 10^{0}C and 30^{0}C, the flowers remain at 30^0-35^{0}C throughout two to four days period. Thermal ability is achieved by an increase in the rate of heat production in proportion to the decrease in ambient temperature: A pattern identical to that in endothermic birds and mammals.

The homoeothermic episode begins with the petal opening and continues through the period of stigma receptivity. It may enhance and stabilize floral development and perhaps reward the insect pollinators with a warm and equable environment. Heat production by flowers is also known to occur in many species of the aroid family and in a few water lilies and palms. Temperature regulation in *Nelumbo* has apparently evolved independently. *Nelumbo* flowers are protogynous and have a variable sequence of anthesis that favours crossing through insect pollinators. A conical receptacle that contains several carpels forms the bottom of an otherwise empty floral chamber before the petals open. Stamens grow around the receptacle, and are initially pressed close to it by the petals, but are released when the petals open fully.

Starch present in the staminal appendages of *Nelumbo* is thought to be the major site of heat production. A common explanation for heat production in flowers is that it enhances evaporation of flower scent, which attracts insects and protects flowers from the cold. This is an incentive for pollinators, particularly endothermic flying insects. Schneider & Buchanan (1980) with several others have concluded that the floral structure and pattern of anthesis of *Nelumbo* were appropriate for the beetle pollination. Beetles, that were trapped overnight, fed and copulated within the floral chamber and were released unharmed the next morning to carry the pollen to other flowers on flower re-opening.

Honey bees on lotus flower doing pollination

Yellow Lotus flower re-opening

Second day bloom of *Nelumbo nucifera* (pink) in a pond

One function of the thermal stability, therefore, may be a direct energy reward to the pollinators. The environment in the floral chamber maintains a high body temperature for the insect and promotes not only feeding, digestion and reproductive behaviour, but also suitable body temperature for flight. Many insects, including beetles and bees, require thoracic temperatures above 30^{0}C to initiate flight and the lotus flower could be directly preparing them for departure, thus eliminating their requirement for endogenous heat production (Seymour & Motel, 1996). For the temperature regulation in sedentary flowers, two reasons can be suggested, why thermo-regulation evolved in some plant species. First it may create a warm and stable environment for the pollinators and thereby facilitate reproduction. Large insects that carry pollen from one flower to another typically require a high body temperature for flight, locomotion and social interaction and they often expend a great deal of energy keeping warm. Those visiting thermogenic flowers are provided with a fairly steady level of heat directly from the plant. Alternatively, the flower itself may require a constant temperature for correct development of its own reproductive structures, or for protecting sensitive parts from damage that might occur if heat production were uncontrolled. The pollen sticks to escaping insects, which fly off to repeat the cycle. This sequence prevents self pollination and promotes the cross-pollination that favours reproductive success, thereby increasing genetic diversity and strong offsprings (Li & Huang, 2009).

LOTUS SEED LONGEVITY

CHAPTER-16

The lotus is known as one of the oldest flowering plants on the planet earth and is the symbol of eternity. The plant is known from geological records as early as 135 million years ago dating from the time when dinosaurs roamed the earth in the Jurassic age, therefore, known as the 'living fossil' along with other relic plants such as *Ginkgo*.

Jane Shen-Miller, a plant physiologist at the University of California, Los Angeles, obtained seven brown oval-shaped lotus seeds from the Beijing Institute of Botany, P.R. China. In 1995 it was reported that scientists caused to germinate a lotus seed which was 1,288 years old. The impervious seed coat and the presence of hardy proteins in the embryo and cotyledons may partly explain the remarkable longevity of the ancient and sacred lotus.

The lotus seed is so robust that it can sprout after centuries of exposure to low-dose gamma radiation. Shen-Miller stated that we must learn about its repair mechanisms and its biochemical, physiological as well as molecular properties. The repair mechanism in the lotus would be very useful if it could be transferred to crops, such as rice, corn and wheat whose seed life span is only a few years.

Activity of the protein-repair enzyme L-isoaspartyl methyltransferase (MT) in the ancient lotus seeds is persistent during germination. The degree of oxidation in the seeds is minimal; racemization of amino acids and changes in the structure of proteins also occur in diseased tissue associated with aging. In the case of lotus seeds, protein degradation (racemization) is minimal. The hard coating of the seeds is another factor which is impervious to water and forms an airtight architecture. The seeds also contain a high level of ascorbic acid (Vitamin-C) and glutathione. These two agents are highly effective in maintaining the life of proteins. Ascorbic acid and glutathione play an important role in protecting the cells from free radical damage: a phenomenon that is believed to be the cause of some of the adverse effects of disease and advancing age. Studies have revealed that lotus seeds contain

Matured seeds of Lotus

substantial amount of polyunsaturated fatty acids which also help in maintaining the longevity of seeds. Glutathione and ascorbic acid as found in lotus seeds may contribute to slower the aging in humans. Xia (2012) observed the effect of germination on the chemical and functional properties of lotus seeds.

During the earlier decades of 20th century, interest was centred largely in the germination of seeds in the *Nelumbo* species/ races. East Indian Lotus seeds obtained by Ohga (1926 a) from a prehistoric peat bed in South Manchuria (China), have been the most striking example of protracted dormancy of seeds of higher plants known to science (Meyer, 1930). These seeds were estimated to be three to four hundred years old. Such a long period of dormancy has not been recorded in the American *Nelumbo* species, but the structure of seeds in the two species is practically alike and no sign of swelling has been observed in either of them after being kept in water for long periods, showing impermeability of the seed coat (Stefferud, 1959; Slocum, 1985).

***Nelumbo nucifera* (Pink single bloom)**

GENETIC AND MOLECULAR IMPROVEMENT

CHAPTER-17

As mentioned above, a lotus is also a popular ornamental, vegetable and medicinal plant, with great potential of utilization in reality, based on which, three types of lotus, named as flower, seed, and rhizome lotus, were defined. Each type of lotus shows notable abundant variable phenotypes, which provide suitable germplasm for its breeding and further study on different traits.

There are nearly 400 *Nelumbo* cultivars available mainly in India, Japan, China and U.S.A. On checking the germplasm, there can be one cultivar with several names or one name for several cultivars. Horto-taxonomical studies on the cultivated plants will avoid this confusion (Sharma & Goel, 2006).

For the genetic improvement of lotus cultivars, the first requisite is wider enrichment of germplasm collection from various phytogeographical regions. Different cultivars collected and introduced from various sources should be grown separately in ponds 2-4 m^2 and 1 metre depth. Ponds should be separated 1 metre apart for carrying out the operations to avoid mixing of the seeds blown from nearby ponds. Cultivars with details are labelled and records are maintained. Vegetative and flowering characters are recorded and cultivars should be photographed for the preparation of their detailed morphological descriptions.

(A)

(B)

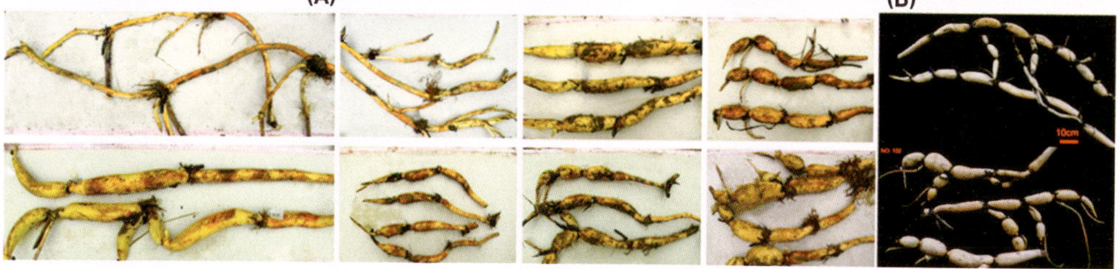

(C)

The diversified phenotypes of the Asian lotus germplasm : (A) Flower lotus germplasm showing different flower color and shape. (B) Lotus seed germplasm showing different size and shape of seed and seedpod. (C) Lotus rhizomes germplasm showing different branching, elongation and expansion of the rhizomes

There are two species of the genus *Nelumbo*: *N. nucifera* Gaertn., the Old World species and *N. lutea* Pers., the New World species. *Nelumbo* is diploid with chromosomes 2x=16. There is no crossing barrier between these two species. Hybrids of these species have greatly helped in enriching the genetic diversity and improvement of the lotus flower. The objectives for genetic improvement of the lotus should be for the need of the market.

Plants should be smaller in size with fewer leaves and profuse blooms. The blooming period now which lasts 3-4 days should be extended. The vase life of lotus cut flowers, especially single and semi double cultivars, should be extended to 4-5 days. The lotus bears a mild fragrance, if it is increased in the flower it will be a welcome step in the improvement of the lotus cultivars. The lotus perfume is a very costly and a rare commodity in the national and international markets.

Pollination in the lotus is carried out by insects mainly honeybees. Different maturity time of stamens and stigma helps to avoid self-pollination. July and August are the best months for cross pollination studies.

Breeding

Recently, a number of studies have been conducted focusing on the genetic and molecular mechanisms underlying the formation of different traits of lotus flower, seed and rhizome. These traits could largely determine the economic value of the lotus, hence becoming the main factors selected in its breeding. Several genetic maps have been constructed through crossing between different germplasms with contrasting phenotypes in some of the economic traits, based on which a number of molecular markers associated with the target traits were developed, including ISSR, AFLP, SSR, RAPD, and SRAP.

Parents with the desired characters are selected from the gene pool. Good seed- bearing lotus cultivars should be selected as the female partner in the breeding programme. Lotus flowers start opening in the early morning so the best time for crossing is between 6.0 and 8.0 A.M. On the first day of flowering when the stigma is receptive, pollen is collected from the male partner (flower of the other cultivar) in a petri dish. With a brush, pollen is spread on the wet stigma. Fertilization process takes six hours duration. In a month's time seeds mature and become ready to be collected. Seeds are stored separately in jars and correctly labelled with all details. Next February, the hybrid seeds are sown and observations are recorded on their vegetative and floral characters. If F1 hybrids show promising results, plants should be preserved. When a seed plant is deemed uniquely outstanding, the first job of the breeder is to multiply the population. Vegetative reproduction through the rhizomes will provide exact duplicates of the original plant. Once a good number of plants are available, the new cultivar can be given a suitable name befitting its phenotypic characters. It is entirely the breeder's choice to name the new cultivar after its most expressive characters or some personality. From the check-list it should be confirmed that the name does not already exist. The varietal name should be registered with the International Plant Society and patented. For the registration of *Nelumbo* and *Nymphaea* cultivars, the International Waterlily and Water Gardening Society (IWGS) is recognized by the International Society for Horticultural Sciences. IWGS is based at Granna Walska, Lotus Land, 695-Ashley Road, Santa Barbara, California, 93 108 (USA). Subsequent plants reproduced vegetatively will carry the same cultivar name. Now this new cultivar can be brought in the floriculture trade.

Plant breeder Perry D. Slocum from U.S.A. has created many attractive and hardy lotus cultivars by crossing the two *Nelumbo* species. Some of his hybrids are unique and stunning in their form and colour. 'Perry's Giant Sunburst' is a cross between

N. nucifera 'Alba Plena' and *N. lutea*, developed in 1987. This cultivar is a summer blooming one with soft yellow flowers, 25-30 cm in diameter and suitable for growing in large containers. Another remarkable cultivar is 'Carolina Queen' developed by him in 1984, which is a hybrid of *N. lutea* x *N. nucifera*. 'Rosea Plena' with a magnificent colour change effect from rose with yellow centre to pink and finally creamy yellow on the third day. 'Angel Wing' is another successful cross made in 1984 involving *N. nucifera* 'Pekimense Rubra' x N. 'Shirokunshi', the tulip lotus. Flowers of this hybrid are delicately white and slightly fragrant, highly suitable for the home garden and recurrent bloom (Green, 2002; Slocum & Robinson, 1996). Induced mutation by irradiating the seeds and rhizomes with gamma rays is another method for improving and developing new cultivars of the lotus. Meanwhile, whole genome re-sequencing on the natural germplasm also identified abundant SNPs and Indels. Together, these data will undoubtedly facilitate the lotus breeding research.

Flower of Lotus

Through breeding and artificial selection, many cultivars with mixed colors have been obtained for the purpose of increasing its ornamental value. A large-scale analysis on the pigment composition of different germplasm has shown that the yellow and red color is mainly determined by the contents of carotenoids and anthocyanins, respectively.

Genome-wide analysis of the *MYB* gene family indicated that there is a similar anthocyanin biosynthesis regulatory system in lotus and *Arabidopsis*, based on which an overexpression of *NnMYB5* in *Arabidopsis* resulted in the accumulation of anthocyanin in immature seeds and flower stalks. In spite of this similarity, a comparative proteomics study between white and red cultivars showed that the expression of the *ANS* gene might be the major reason for the absence of anthocyanin biosynthesis in the white flower lotus. Further analysis found that different levels of methylation occur in the promoter regions of *ANS* gene between the two cultivars, which indicates the epigenetic regulation on expression of this gene. However, the genes that lead to the different methylation level on the promoter of *ANS* gene between the red and white lotus cultivars is still unknown. In addition, there are cultivars showing genetic constant spotted color, which is still not understood. It will be very important not only to the breeding of flower lotus, but also to enriching our knowledge on the coloration of plant flowers to explore the mechanism underlying the regulation of spotted color in lotus.

Lotus cultivar with spotted color flower

In addition to color, flower shape is also important for the economic value of ornamental plants. Based on different purposes of breeding, lotus cultivars with diversified flower shapes were obtained, including few-petalled, semidouble-petalled, double-petalled, duplicate-petalled and all-double-petalled cultivars. For the semidouble-petalled, double-petalled shapes, they are usually the resultants of stamen petaloid. Comparative transcriptomic studies among petal, stamen petaloid and stamen through RNA-seq were conducted, which identified several candidate genes involved in stamen petaloid, especially some MADS-box genes [58]. Their study revealed 11 MADS-box genes and one *APETALA2* (*AP2*) gene being involved in the stamen petaloid phenomenon. Among them, *AGL15*, *AGL80* and *AGAMOUS* genes are positively related to, and *AGL6* is negatively related to the stamen petaloid. Meanwhile, a genome-wide DNA methylation analysis was also conducted among these three tissues, which indicates the potential involvement of epigenetic regulation on the stamen petaloid. However, this study did not detect any obvious changes of the methylation on the MADS-box genes. There also exist pistil petaloid cultivars, in which the stamen petaloid also occurs. How these are coordinately regulated is still unknown in the lotus. Furthermore, it is well known that lotus bloom in the summer days, which brings some challenges for its wide utilization in ornamentation. It will be very important to make it bloom either earlier or later for ornamental purposes. Hence, unveiling the mechanism controlling the time of flowering is also important. A transcriptomic analysis has been conducted aiming at exploring the candidate genes that control the time of flowering, which indicate the existence of a complicated regulatory network. Their data indicate that the differential regulation of some photoperiod related genes, such as *COP1, CCA1, LHY, CO-LIKE*, and *FT*, the vernalization gene *VIN3* and the gibberellic acid-related gene *GAI*, might be involved in the regulation of early flowering in lotus. Specifically, several isoforms of the *FT* gene were found to be differentially expressed.

Rhizome and Seeds

As mentioned above, lotus is not only an ornamental plant, but also a vegetable because of its edible rhizomes and seeds. Lotus has a morphologically modified subterraneous stem. Especially for the temperate ecotype, its subterraneous stem is enlarged in autumn, which is known as rhizome. The rhizome contains abundant starch, proteins and vitamins, making it a popular edible vegetable. Enlargement of lotus rhizome could largely determine its economic value. In addition, the enlarged rhizome could also help the lotus to survive from winter during its bud dormancy, and provide substrates and energy for its asexual propagation. This phenomenon is very similar with the tuberization of the potato, which has been proven to be regulated through a very intricate genetic network. Being a significant feature distinguishing between the temperate and tropical lotus, it may also facilitate in understanding the evolution and domestication of the lotus. It seems that rhizome enlargement is tightly related to the flowering in a

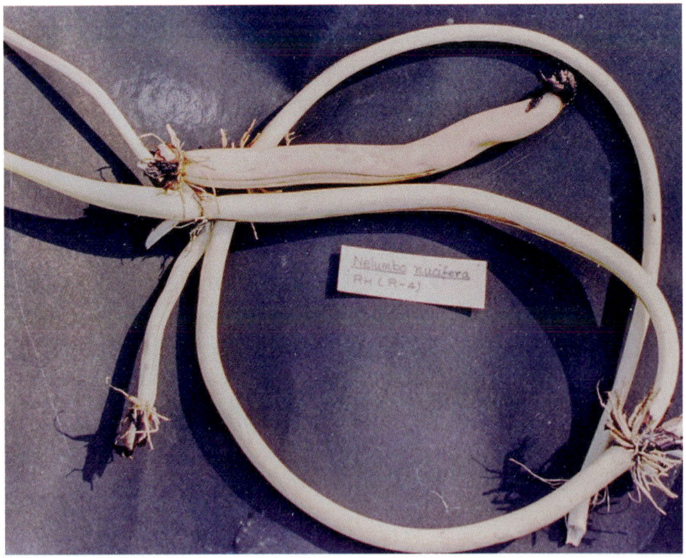

Young rhizomes of Lotus

Lotus matured seeds

lotus. Usually, the enlargement occurs after flowering. For the purposes of increasing its yield in agricultural production, genetic and transcriptomic studies focusing on the enlargement of this rhizome have been conducted.

Gene expressions during the rhizome development were analyzed through RNA-Seq, which identified the specific candidate genes for rhizome enlargement. The results also indicated the role of SNPs and alternative splicing (AS) events in Asian lotus rhizome development. Similar with the yield traits in many crops, the enlargement of the lotus rhizome is a quantitative trait. Developing a suitable genetic population and constructing high density genetic map will be very helpful to elucidate the mechanism underlying rhizome development and enlargement.

Besides its longevity, lotus seed is also edible either fresh or dry matured, with an additional medicinal versatility resulting from compounds like alkaloids, flavonoids and certain micronutrients. Both the size and number of the seeds per seedpod vary among different lotus cultivars. It is very important to increase its nutrition as well as its yield in lotus seed production. To achieve this, comparative proteomics and metabolomics studies were conducted on lotus seeds during its development, which not only deepen the understanding on the development of lotus seed, but also determine candidate genes crucial for lotus seed size. In addition, comparative transcriptomic analysis was also conducted between two lotus germplasms with contrasting phenotypes in both seed size and seed number per seedpod (Torus). Similar to rhizome, the yield of seed is also a quantitative trait, which requires more study at the genetic aspect. Meanwhile, because of its medicinal usage, it is necessary to conduct a comprehensive analysis on it metabolites during seed development.

Studies on the Establishment of Lotus Regeneration and Transformation System

To be a model horticultural plant, it might be necessary to establish a transformation system, which will facilitate the studies on the functions of different genes in the lotus. A study was conducted to induce the formation of a callus from different explants of the lotus, in which somatic embryo cultivated in suitable medium containing a combination of different growth regulators was proposed. To obtain more in-depth understanding, a proteomic analysis was conducted to identify the key proteins that might be critical for the induction of callus from developing cotyledon.

Directly inducing the formation of a shoot from the bud has also been successfully performed. Based on this system, various studies have been conducted to transform the lotus. It seems that the induced shoot from the embryo apical bud could be successfully transformed through a particle bombardment device with a pCAMBIA2301 vector. This method not only succeeded with the GUS reporter gene, but also with the anti-sense of two anthocyanin biosynthesis genes *dihydroflavonol 4-reductase* (anti-*DFR*) and *Chalcone synthase* (anti-*CHS*). Except for the group from Thailand, there are still no other studies conducted successfully on the transformation of the lotus, although a lot of researchers are working on this. It seems there are still challenges on the reproducibility and the efficiency of the transformation, as well as the selection of a suitable cultivar.

CONSERVATION OF LOTUS

CHAPTER-18

Worldwide attention for the protection and conservation of the germplasm resources of aquatic plant species and their habitats has constantly been of great concern to scientists all over the world (Heijny, 1978; Schneider & Kodono, 1984; Kachroo, 1984). The aquatic bodies and wetlands comprise the most important components of natural ecosystems. During the last six decades, aquatic habitats have been deteriorating rapidly due to extensive demographic pressure, urbanization and invasion of aquatic weeds (Varshney & Singh, 1976). Increased inflow of industrial effluents causes environmental degradation which is a serious threat to the natural ecosystems and aquatic habitats. Soil washed down from denuded slopes and fields is swiftly choking the wetlands and rivers. As a result, fragile ecosystems, such as tropical moist forests, lakes reservoirs, mangroves, wetlands and coastal regions which constitute the significant hotspots of biodiversity, are becoming endangered. Global warming is among one of the root causes destabilizing the equilibrium of wetland ecosystems (Babu & Gopinathan, 1986).

The Indian subcontinent harbours approximately 14,50,800 hectares of natural wetland ecosystems which are rich in aquatic flora. With about 470 species of flowering plants, this subcontinent possesses approximately half the aquatic angiospermic taxa in the world, representing 59 families (Lavania *et al.,* 1990). India ranks second only to tropical South America in the number of endemic aquatic plant species. Aquatic habitats are found scattered all over the country and are independent of climate and altitude. The aquatic plant species, perhaps with the exception of few beautiful and attractive flowering ones, have received little attention during the last few decades.

Despite recognizing the life-sustaining importance of ponds, lakes, rivers and marshes with aquatic plant diversity, they have also been exploited for various purposes such as water sports and for the disposal and discharge of all kinds of waste as well as industrial effluents comprising toxic substances without

Pond in CSIR-NBRI Botanic Garden, Lucknow full of Lotus flowers

Various stages of flowering and fruiting in *Nelumbo nucifera* (pink) in a pond

understanding their immense destructive potential. The macrophytes growing in aquatic habitats are not recognized as important components of such delicate ecosystems. It is only during the last few years that the importance of these life forms has been realized and the conservation of aquatic plant habitats has been initiated for their sustainable use. Aquatic plant species perform several functions in their natural ecosystems. The submerged aquatic taxa oxygenate the water and serve as a source of food for animals as well as human beings.

So far the systematics of Indian aquatic plants has not been given due attention and even morphological studies are so scarce that these taxa are generally misidentified. Aquatic plant species exhibit a higher degree of morphological variability and their vegetative characteristics are very deceptive. They are known for their unusual phenotypic plasticity which may be an adaptive response to their dynamic environment (Subramanyam, 1962).

Plant species growing in aquatic ecosystems adapt themselves to the specific habitat conditions and cover little surface area. They also comprise a much smaller number of taxa per unit area than the terrestrial ecosystems (Jain, 1990). The natural aging and eutrophication of water bodies also have adverse effects on the aquatic vegetation. The threats to aquatic vegetation fall broadly under the following four main categories (Jain, 1990).

- ✓ Direct Threats
- ✓ Habitat alteration
- ✓ Habitat destruction
- ✓ Population depletion

All these threats exist actively throughout the Indian subcontinent. It has been a common practice in India to remove massive vegetation from ponds and lakes for making water bodies available for the cultivation of water chestnut, fish culture and for water sports. Lotus and Makhana fall into the fourth category and their over-exploitation for various purposes is common practice all over the country. This has resulted in the considerable depletion of their population. The disappearance of Makhana (*Euryale ferox* Salisb.) from the Dal Lake in Srinagar, Kashmir, due to pollution from domestic refuse, sewage and field outwash has been quite frequent during last two decades.

A virtual root-out of the genera *Euryale, Nelumbo, Nymphaea* species, is reported from North Bihar due to recurring floods (Jha *et al.*, 1991). Kaul (1984) and Kak (1985) have strongly advocated the need for conservation of aquatic plant habitats in North -Western India. The common practice of discharging domestic and industrial effluents into the aquatic ecosystem has caused drastic changes in the flora and fauna of the wetlands of our country. Due to the endangerment of their habitats as discussed above, the population of *Nelumbo* and *Euryale* has started shrinking rapidly.

In the Manipur valley lotus growing area is nearly 55 hectares. Systematic cultivation of the lotus is not only good to improve the local economy of the people but also helps the conservation of the wetlands. In Chhatisgarh rich natives prepare small ponds for planting the lotus. In home gardens lotus is the essential component to bring happiness and prosperity.

Ex-situ Conservation of *Nelumbo nucifera* Gaertn. in CSIR-NBRI

For the *ex-situ* conservation, species/races of *Nelumbo* were collected, acclimatized and multiplied in the aquatic bodies of the Botanic Garden, CSIR-NBRI, Lucknow. The germplasm collection of 35 indigenous races was builtup by organizing plant collection tours to different phytogeographical regions of the country as well as through the exchange of plant material with the various organizations and the Botanic Gardens. The indigenous germplasm collection of *Nelumbo* species/ races was procured from eleven states: Bihar, Chandigarh, Karnataka, Madhya Pradesh, Maharashtra, New Delhi, Odisha, Puducherry, Tamil Nadu, Uttar Pradesh and West Bengal (Sharma & Goel, 1997; Goel, 2000).

The germplasm of 25 races was enriched by procuring the plant material from USA, Germany, Japan, UK, Thailand, Brazil and Australia in various shades of pink, white and yellow flowering species/races of *Nelumbo*. These races were acclimatized, documented and multiplied in the Botanic Garden, Lucknow.

***Nelumbo* sp. in pink-white flower**

***Nelumbo lutea*- Yellow flowered lotus**

For studying the diversity of *Nelumbo nucifera* races, the observations on morphological characters: shape, size and colour of rhizome, colour of petiole and pedicel, colour and shape of prickles; shape, size colour and venation pattern of leaf; colour and size of bud and flower; size, shape, number and colour of sepals, petals, torus, stamens, stigma and seeds were taken into consideration in the pink and white races. The maximum number of petals ranging from 116-160 was recorded in a pink double race from Midnapur (West Bengal) (Sharma *et al.,* 1998).

The germplasm material of *Nelumbo* species/races, which was acclimatized and multiplied successfully in the climatic conditions at Lucknow, was disseminated to various, institutions, connoisseurs and progressive farmers for promoting the lotus culture (Sharma & Goel, 1997, 2000, 2001; Goel *et al.*, 2001).

PROPAGATION OF LOTUS

CHAPTER-19

The lotus is sexually propagated by sowing the seeds and asexually by planting the rhizomes.

Seed Propagation

Lotus seed is impervious to water and it can remain viable for several years. Studies at the Royal Botanic Gardens, Kew (UK) have revealed that the seeds collected from over 150 years old herbarium specimens were found to be viable and germinated successfully (Hong *et al.,* 1998). To overcome that dormancy, Ohga (1926) found concentrated sulphuric acid treatment as an easy method for better germination of the lotus seeds. However 50% chromic acid and concentrated potassium hydrate also help in overcoming the impermeability after a longer period of time. It has been found that the cellulose layer of cells (tissue) participates in maintaining the impermeability of the seed coat to water and may be chiefly responsible for the dormancy in the seeds of *Nelumbo* (Jones, 1928). But scarification of seeds was found to be the easiest and safest method for raising lotus seedlings.

For practical purposes the storage of seeds for eight years has no effect on the germination rate. For one hectare plantation of the lotus crop 10 kg seeds are required. The hard coat of the seed should be scarified at both ends. Care should be taken not to damage the flesh of the seed while filing. Seeds are soaked in warm water (25-30^{0}C) with twelve hours of light to encourage germination. Water should be changed daily until the seeds germinate which take from five to eight days. There are enough nutrients in the *Nelumbo* seed to support the plant for thirty days, but it should be transplanted after 3-4 weeks when seedlings have two or three leaves and a few roots. Seedlings should be planted separately in individual pots and placed in water. The depth of the water should be increased gradually as the plant grows. Plants do not like root disturbances and should be transplanted into their permanent positions as soon as possible. Late in spring rolled-up leaves push out of the mud. The crinkled leaf slowly opens during the warm days and soon becomes a big platter with a depression in the centre where the petiole joins it from below.

Lotus flower, buds and torus

There is a problem with seed-grown plants. Seeds are highly heterozygous therefore the progeny may not be true to the original variety. Seeds are the combination of genetic material from two separate plants which is the result of pollen from one plant introduced to the stigma of another plant by human, bee or other pollinators. The resultant offspring will differ from both parents. Furthermore, germinated seedlings will not produce a crop until the following season.

Vegetative Propagation

Healthy rhizomes are selected for raising the lotus crop. To ensure blooms in the first season, rhizomes with at least two growing points should be planted. The growth points must be protected, if they are broken during plantation, the entire rhizome will die. They are shallowly planted in clay soil with rich organic matter. Rhizomes are planted at a 30^0 angle in the mud. The distal end should be placed 10 cm inside the mud and the proximal end should protrude from the water. The crown of the rhizome should not be covered with soil.

Initially the depth of water should be only 6 cm, raised gradually up to 30 cm as the crop grows. The ideal water temperature for rhizome plantation is 25-27^{0}C. The growing tip should be pointing towards the centre of the pond or pot, as this will develop into a runner. This will branch off and spread throughout the mud, producing aerial leaves and flowers as the season progresses (Goel, *et al.*, 2003).

The rhizome grows at a surprising rate of about ten metres in a year. At each node as the point of growth, the plant produces first a leaf scale on the lower side, then another one on the upper side, followed immediately by a foliage leaf. From the axil of the upper leaf scale a flower emerges, which gives the plant great potential for flowering and spreading at the same time.

Tissue Culture

Mass multiplication of lotus plantlets through the tissue culture method of taking ex-plant from shoot tips was reported by Liang and Zhao (1998). They studied the effect of sucrose concentration, ex-plant size and subculture cycle on propagation co-efficient, and different planting dates on rhizome production. Embryos are also used as explants. However, tissue culture technique is not a common practice due to easy vegetative propagation.

Germination studies of Lotus in Cement tubs

Lotus rhizomes (vegetative propagation) for sowing

Tissue culture taking ex-plant from shoot tips in Lotus

PESTS AND DISEASES

CHAPTER-20

Pests

Aphids: They appear on the lotus plants when the temperature is on the rise. Aphids appear on the petioles, leaves and flower buds and multiply very fast. When the pest attack is serious, young leaves wilt and flower buds become dry. To protect lotus plants from these pests, kill them manually at the beginning. Spraying with a strong jet of water will knock down the aphids into water where they will be eaten by the fish. If it is more serious, pesticides should be sprayed on the plants.

Beetles: This is a common pest on lotus plants. The beetles may either be picked off by hand or dislodged by spraying a jet of water.

Snails and Shrimps: They cause great harm to the lotus plants. They multiply very fast and create problems in the cultivation of the lotus. For controlling them, hand picking and killing is the most effective measure.

Diseases

Dry Spot Disease: The disease appears on the leaves. It is caused by the fungus *Phyllosticta hydrophila*. The yellow spots on leaves later turn to brown patches. It seriously impairs the growth of lotus plants.

Brown Strip Disease: This disease is caused by the fungus *Alternaria nelumbii*. Standing lotus leaves are badly affected. Yellow spots are observed on the leaves which later on turn to dark brown patches. Fungicides are used to control the disease.

Putrescence Disease: The disease is caused by *Fusarium oxysporum*. Infected plants bear light green leaves with yellow edges. Gradually leaves turn brown and dry completely. Later on the petioles bend downwards and the leaves die. The fungus can spread to newly formed rhizomes in the form of mycelium and spores. As a preventive measure, the soil mixture should be sterilized before plantation for controlling the disease.

Aphids on upper surface of lotus leaf

Weeds

Aquatic weeds are sometimes a major problem in lotus cultivation particularly during the summer months when the growth of weeds is faster. Commonly found weeds in lotus pond are water-hyacinth (*Eichhornia crassipes*) which spreads fast and covers the entire water surface. Species of *Typha*, *Lemna, Wolfia*, are the other common weeds found in lotus ponds. Freshwater algae, mainly *Spirogyra*, also cause much damage to the crop. *Ceratophyllum demersum, Potamegaton crispus* and *Salvinia natans* are also problematic weeds which are found in a lotus pond. Mechanical removal of the weeds is a safe and easy method. However, with a severe weed problem under large growing areas, suitable herbicides may be applied to control them.

Lotus with Pink flower

Dorsal view of a leaf with dry spot disease in Lotus

LOTUS CULTIVATION

CHAPTER-21

Lotus cultivation is a great pleasure and a profitable business. It is an aquatic crop which is grown in shallow water ponds. All lotuses are day-bloomers, open in early morning and close in the afternoon for three to four successive days.

Season

The lotus is a summer growing crop (March-September). It is a long-day plant and flourishes in full sunshine. Lotus starts aging from October onwards. It is planted in February (spring) when the average temperature is 20^0-25^0C. But it can also be planted in March. It is advisable to plant it earlier so that plants get sufficient time for developing flowers and rhizomes before dying back or dormat in water body.

A view of Lotus pond

Soil Preparation

The lotus prefers to be grown in an organically abundant soil medium. Soil mixtures are prepared with 2/3 loamy clay soil and

A view of a aquatic body

1/3 of farmyard manure. Low Phosphorous, high Nitrogen and Potassium are beneficial to the plant growth. The soil mixture should be 20-30 cm thick at the bottom of the pond. The loamy clay mixture helps to anchor the rhizomes as they develop.

Water

The lotus grows in a wide range of pH but prefers fresh water. It can grow in a pond having pH 6.0-6.5 and 7.5-7.8. The optimum pH 7.5 is ideal for the cultivation of the lotus. The depth of the water should be 0.5-1.0 m.

Seed Requirement

The average weight of the lotus seed is 1.0 gm. Thus 10 kg seeds are required for one hectare of pond area.

Rhizome Requirement

For one hectare plantation of the lotus crop, 50 kg rhizomes are required. The density for planting of rhizomes is recommended at 2 x 2 m apart which constitutes 2,500 rhizome segments.

The lotus can be propagated vegetatively by the division of rhizomes and also through seeds. The rhizomes with new sprouts are cut into small pieces with three nodes each and planted horizontally at the depth of 10 cm, keeping the eyes above the soil from February to April. The rhizomes can produce large flowering plants in the same season whereas the seedlings raised from seeds require at least two or three years to produce flowers under subtropical conditions.

The propagation through seeds is practiced in the months of February and March. The seeds are scarified with sand paper at both ends and immediately immersed in water. The germination takes place after five to seven days at normal room temperature. The seedlings are ready for transplanting to the aquatic bodies after 15-20 days. The seedling growth speeds up as there is a rise in the day temperature (30^0-35^0C). *Nelumbo nucifera* mainly hails from the tropical and sub-tropical regions, so it requires a warm and humid climate for best performance. It is easily cultivated in ponds and lakes with one metre depth of water.

The bottom of the aquatic body is filled with a mixture of clay soil and farmyard manure in a 3:1 ratio with 100 gm each of neem cake and DAP per sq m in a 0.5 m thick layer. The transplantation of rhizomes or seedlings is generally done during February and March. The water level is gradually increased according to the growth of the plants in the aquatic body. The flowering in the lotus starts from April and continues till September-October in the climatic conditions of the North Indian plains. Lotus plants become dormant during November onwards until January due to the low temperature in the winter season. Optimum growth of seedlings under laboratory conditions was recorded at 30^0C, while it was minimal at 20^0C and below. It has also been observed that the seeds could also germinate under high alkaline (pH 9.0) and acidic (pH 5.0) conditions with a low germination percentage. However, extreme alkaline or acidic conditions lower the overall growth of the seedlings. The optimum growth of plants was recorded in the pH 7.5. The effect of burial by sedimentation on the germination of seeds was significantly higher than with the seeds buried at less than 2.0 cm depth, or on seeds not buried at all. The germination percentage of seeds reduced significantly up to 60% when they were buried under 2 cm of soil (Nohara & Tsuchiya, 1990).

Lotus plants require full sunshine and warmth for the ripening of the rhizomes. After attaining maturity the chain-like cylindrical rhizomes develop additional anchor and feeder roots, which penetrate the clay soil quite deeply. They are extremely difficult to uproot from lakes and ponds. Different races should not be grown together in one pond. Their germplasm should be maintained separately in different chambers or separate aquatic bodies.

The growing points of rhizomes are very delicate and great care should be taken while handling and transplanting them in the ground. These points should not be damaged otherwise the rest of the root stock will rot in their absence. The frost and hailstorms can also damage the leaves and flowers of *Nelumbo* during the growing season when cultivated under open conditions. Violent winds can also make a sorry sight of tender lotus leaves and flowers. Lotus plants grow profusely during the summer and rainy seasons. The flowers are produced on long pedicels above the level of the leaves. The flower buds are picked at maturity and packed two or three days before opening so that they are able to withstand long journeys.

Growth

Rhizomes of *Nelumbo nucifera* can grow upto 4-6 metres in an ideal season. The first growth phase is vegetative, in which each leaf petiole that emerges from a node is larger than the previous one. When new petioles are shorter than the older ones it shows that the plant has switched to producing storage rhizomes. Removal of leaves may also improve the rhizome quality significantly. Higher yields are obtained if the flowers are also removed (Yamaguchi, 1990).

Hall and Penfound (1944) stated that the rhizomes of *Nelumbo lutea* growing in pure sand extend radially to an average of 15 metres in one growing season. This represents a growth rate of 7.0 cm per day for the entire summer. On this basis six properly spaced patches of lotus rhizomes (3 m across) would produce an acre of lotus crop during one growing period.

Harvesting of Rhizomes

The rhizomes are ready for digging in October. Rhizomes are harvested at about 120 days in warm climates and 150-180 days or after the leaves have died in sub-temperate climates. Seeds are harvested after the pod turns brown. Traditionally lotus rhizomes are harvested by hand when the seeds mature at the end of summer. At the end of each growing season about 20% of the rhizomes should be left in the pond for the next season. It is suggested that a separate pond be used for propagation for the following year. Rhizomes are frost-resistant as they are protected by mud and water.

Generally there are two types of rhizomes, white and buff-orange coloured. The farinaceous rhizomes of the lotus plant are collected and sold as vegetables in the market. They measure 60-120 cm in length and 4-8 cm in diameter. In cross section rhizomes show a few large cavities surrounded by several small pores. They are fleshy and when freshly cut, exude a mucilaginous milky juice. They are somewhat fibrous and do not soften even after prolonged boiling.

Yield

The Japanese yield of lotus rhizomes is around 13.7 t/ha, though Japanese yields of 3.5-4.5 t/ha are also reported (Yamaguchi, 1990). Malaysian and Indian yields of lotus rhizomes are much lower, reported at 3.1-8.2 and 3.8-4.7 t/ha respectively (Ong, 1996).

In the Indian market lotus rhizomes are sold at around Rs. 20-60 per kg, flowers at Rs. 5.00 to 10.00 each and seeds price range from Rs. 50 to 100 per kg. Progressive farmers can earn a net profit of Rs.40,000-50,000 per hectare from the lotus cultivation.

Stages of budding, flowering and fruiting in Lotus (pink) in a lake

***Nelumbo nucifera* (pink) in flowering and matured torus in a pond**

The shelf life of the lotus rhizome is two weeks. For the domestic market, lotus should be transported by refrigerated truck. For export, air freight is ideal, but cost should be considered in relation to the market price for the sale of rhizomes.

Market Quality of Rhizomes

Rhizome skin colour	–	Milky white
Rhizome size	–	Three segments with diameter more than 4 cm
Rhizome flesh	–	High water concentration of flesh with a soft and crunchy texture.
Packaging	–	Rhizomes should be packed in cartons of 1, 2, 4, 5 and/or 10 Kg (Nguyen 1998)

Demand and Market

There is a good demand for fresh rhizomes, dried, tinned and also as the frozen products. Lotus flowers and pods are traded in the floral market. Boiled lotus rhizome is an essential ingredient for a Japanese vegetable salad. Leaves are used as wrappers. Lotus is popular as a crunchy ingredient in Japanese tempura style cooking. There is a good market for salted rhizomes in Japan, China, Taiwan, Hong Kong and Singapore. Lotus is a potential crop for India but its systematic cultivation is yet to be popularized.

MEDICINAL IMPORTANCE OF LOTUS

CHAPTER-22

Nelumbo nucifera the sacred lotus has been valuable for us to using as vegetable, functional food and herb medicine for over 2,000 years. The purpose of this book is to review the traditional/ modern uses, chemical compositions and pharmacological activities of different parts of lotus. Traditionally, this plant has been used to treat chronic dyspepsia, hematuria, insomnia, nervous disorders, cardiovascular diseases, and hyperlipidemia. phytochemical investigations on different parts of lotus have indicated a wide spectrum of at least 255 constituents belonging to different chemical groups, including proteins, amino acids, polysaccharides, starch, flavonoids, alkaloids, essential oils, triterpenoids, steroids and glycosides. Meanwhile, its pharmacological activities, including anti-obesity, antioxidant, anti-inflammatory, cardiovascular, hepatoprotective, hypoglycemic, hypolipidemic, antitumor, memory-improving and antiviral activities, have also been reviewed, together with its applications in health food industry and clinic uses of its single plant or herbal formulation. Herein, this review will provide state-of-the-art overview of its traditional and modern uses in food industry and medicines, together with the comprehensive profiles of phytochemicals and health promoting bioactivities of this valuable plant. In addition, the new perspectives and future challenges in the research on lotus are also outlined (ICMR 2017; Chen, *et al.*, 2019).

At one time, nearly all medicines were derived from biological resources. Even today these resources remain vital and as much as 65% to 70% of modern medicines are derived from natural products (State of the Environment Report, 2001). Use of

A varient of *Nelumbo nucifera* (pink sinlgle flower)

A bud of *Nelumbo nucifera*

A matured and dried torus of *Nelumbo nucifera*

plant based drugs and chemicals for curing ailments and personal adornment is as old as human civilization. Nearly 80% of the world's population relies on traditional medicines for primary health care, most of which involves the use of plant extracts (Sandhya *et al.,* 2006).

Old Indian and Chinese references indicate the powerful medicinal properties of the lotus (*Nelumbo nucifera*) against a number of human ailments involving the digestive, reproductive, circulatory and excretory systems. Almost all parts of the lotus plant have tonic and astringent properties and are recommended for the treatment of spermatorrhoea, gonorrhoea, insomnia, metrorrhagia and dyspepsia (Caius, 1939; Nadkarni, 1954). Phytochemical and therapeutic profile of *Nelumbo nucifera* has been discussed in detail by Mukherjee *et al.* (2010).

In Sri Lanka, an aqueous extract of the fresh rhizomes of the white flowered lotus variety is given orally in doses of 1-4 ounces in the treatment of snake bite. A paste of rhizomes is applied to ringworm and other cutaneous infections. A kind of arrowroot is prepared from the fleshy rhizomes which is aromatic and sweet in taste and is said to be not only nutritious but a tonic and of value in pediatric medicine. It is given to children in cases of diarrhoea, dysentery and dyspepsia. It also forms the basis of the diet for infants. Rhizomes of the white variety of Lotus are credited with the property of restoring and rejuvenating health and are also considered haemostatic.

The large leaves are used as cool bed sheets in high fever with heat and burning of the skin. The milky viscoid juice of the leaf is a useful medicine in diarrhoea, decocted with liquorice for sunstroke and vertigo. Lotus leaf has become popular for lowering blood lipids and treating fatty liver. It is commonly combined with *Crataegus*, which promotes blood circulation and lowers blood fats.

Lotus flowers are used as astringent in diarrhoea, cholera, fever and liver diseases. They are also recommended as cardiotonic. Flowers are decocted for abdominal cramps, blood discharges, metrorrhagia and non-expulsion of the amniotic sac. Syrup of dried flowers in doses of half an ounce is found to be efficacious in curing mild dysentery. In Sri Lanka and South India, the flowers are considered diuretic and cooling. Lotus flowers are boiled in milk with an equal amount of flowers of 'Jamun' (*Eugenia jambolana)* made into a paste and then balls of areca nut size. They are given to women for the relief of false pains in the early months of pregnancy. The dried pinky red petals are used by the Chinese as a cosmetic application on the face for improving the complexion.

The stamens are regarded as astringent and cooling, useful in a burning sensation of the body, bleeding piles and menorrhagia. In bleeding piles they are administered with honey and fresh butter or sugar. The carefully dried, yellow fragrant stamens are astringent and valued as a commended diuretic remedy in China. They are also used for cosmetic purposes. They are reported to purify the heart, permeate the kidneys, strengthen virility, blacken hair and check hemorrhages and spermatorrhoea (Caius, 1939; Mitra & Kapoor, 1976).

Dry stamens of Lotus used for Medicine

Receptive pistils in a young torus of *Nelumbo nucifera*

White double Lotus flower with a bud

Nelumbo receptacle is used as antihaemorrhagic for excess and irregular menstrual bleeding. It is also a remedy for dehydration caused by diarrhoea in summers and used for prevention of miscarriage in traditional Chinese medicines. Later on it has been investigated that the antihaemorrhagic activity of nelumbin receptaculum extracted from the receptacle is due to the presence of quercetin.

The seeds are used to check vomiting and are given to the children as a diuretic and refrigerant. They form a cooling medicine for cutaneous diseases and leprosy and are considered an antidote to poison. They are regarded as a general tonic, recommended for diarrhoea, insomnia, metrorrhagia, nightmares, spermatorrhoea, leucorrhoea, seminal emissions and are considered nourishing and highly useful in preserving general health and strength and in promoting blood circulation and virility.

In China, plumule, caulicle or the geminating embryo of the ripe seeds are given to reduce high fever and said to be useful in the treatment of cholera, haemoptysis and spermatorrhoea. Lotus plumule contains various alkaloids and is used to remove heat or as a tranquilizer and antihypertensive agent in folk medicine. Lotus flowers contain alkaloids, organic acids, amino acids and β-Caratenoid and are used to warm kidney and spleen or as a cardiotonic in folk remedies. To access the possible antioxidant activities of methanol extracts of lotus plumule and flowers, several *in vitro* assays were conducted. It was found that both lotus plumule and flowers possess strong reducing powers and free radical scavenging abilities. Cotyledons are believed to promote virility and recommended to cure leucorrhoea and gonorrhoea.

The whole plant is said to be refrigerant in smallpox and is reported to stop eruptions, an antidote to alcohol and mushroom poisoning and anti-cancer (Poterfield 1941; Hsu 1945; Nadkarni 1954; Chopra *et al.*, 1956; Chopra *et al.*, 1958; Khan 1958; Toyoda 1960; Nakaoiki *et al.*, 1961; Rahman *et al.*, 1962; Kunitomo *et al.*, 1964; Perry 1989; Boyd 1968; Trivedi *et al.*, 1980 & 1982).

The botanical and pharmacognostic studies reveal that there is no difference in the structure whether the rhizome is from the red flowered variety or from the white flowered variety of the lotus plant. Lotus rhizome nodes are astringent and neutral benefiting the liver, lungs and stomach. They are mostly used to control bleeding. Sterol reducing sugar and alkaloids are found in the rhizomes. All parts of the lotus have some antihemorrhagic effect but the rhizome nodes are more relied upon specifically for this purpose.

The leaves of *N. nucifera* have been reported to contain alkaloid nuciferin (Arthur & Cheving, 1959) and a glycoside of quercetin (Nakaoki *et al.,* 1961). Glycoside kaempferol is reported in the flowers (Rahman *et al.,* 1962). A concentrate from an ether extract of petals has yielded quercetin and luteolin while the ether extract of torus yields only quercetin. A concentrate from a water extract of petals and stamens yields isoquercetin and glucoluteolin while the concentrate from the water extract of the torus yields isoquercetin only. Leaves contain both isoquercetin and glucoluteolin. Both pink and white races of *N. nucifera* contain anthoxanthin pigments in different parts of the flower (Nagarajun, *et al.* 1966).

'Kamal' is a well known herb in Chattisgarh. For generations its different plant parts are in use as folk medicines. Infestations of *Nymphula* caterpillars happen not only in Chhattisgarh but also in other parts of India. The caterpillars have tracheal gills and feed on lotus leaves. The traditional healers of Chhattisgarh plains are aware of its medicinal properties and uses. Through recently conducted ethno-entomological surveys, much valuable information has been collected on this aspect. The fully fed caterpillars are used in powder form. Pankaj Oudhia states that it is commonly used internally as medicine alone or in combination with other medicinal herbs. The traditional healers of the Raipur region use the caterpillar powder in the treatment of common fevers. In combination with honey, it is given internally. According to the healers, its internal use helps to reduce the temperature. Caterpillar powder is also used for the treatment of diarrhoea.

Tribal people of the Mayurbhanj district, Odisha, use *Nelumbo nucifera* for curing gynecological disorders. For white discharge, a decoction of rhizomes of the white flowered lotus plant of about 15 ml is taken by women on an empty stomach for fifteen days as a cure (Behera, 2006).

A woman enjoying scent of a lotus flower

Petals used for manufacturing Lotus perfume

LOTUS PERFUME

CHAPTER-23

The lotus is the most admired of all plants with its graceful flowers and delightful perfume. As a flower essence, the lotus has a varied usage. It is called the spiritual elixir that helps in meditation by calming the mind and improving concentration. Lotus is an excellent elixir for balancing, cleansing and strengthening the aura. All the chakras are aligned and balanced by releasing, adding or directing energies to them, thus releasing better health and harmony. Lotus essence can be used in bath therapy. It can be added to an aromatic bath or a massage blend, or diffused in the air to recharge energies and clear away unwanted thoughts. As lotus oil is very intense, it is at its best when diluted or blended with lighter essential oils. Aroma therapists state that lotus essence has always been highly regarded as an elixir for any ailment with a positive effect on the entire body. Lotus is also believed to heighten creative ability and mind power and induce a sense of ease and peace.

The essential oil of *Nelumbo nucifera* contains mainly of myristic acid 10.95%, edulan 8.76% and phytol 5.16%, terpenoidal compounds 9% and hydrocarbons 31.89%. A glycoside extracted from the flower petals has been identified as kaempferol.

According to Christopher McMohan, 20,000-25,000 lotus blossoms make one kilo of concrete and 75,000-100,000 blossoms are required to make one kilo of absolute. The extremely waxy blossoms yield just a tiny amount of absolute.

A matured Lotus bud and flower for perfume

A pinkish white lotus bloom

The petals are filled in the extracting unit and washed with hexane several times to remove the waxes, pigments and essential oil from the flowers. After going through the delicate process of removing a large proportion of hexane from the concrete, it is washed in alcohol. Chilling, filtering and vacuum distilling of the alcohol from the absolute then must take place before one gets the final product or absolute.

The lotus perfume available in the trade is a blend of patchouli, benzoin and styrax with phenylethyl and cinnamic alcohols (Swarup, 1989).

Tight & matured bud of *Nelumbo nucifera* (pink) suitable for perfume

ECONOMIC IMPORTANCE OF LOTUS

CHAPTER-24

Almost all parts of the lotus plant are edible and eaten in Asia and Australia. It has been reported that rhizomes of lotus were used as food during the Deccan famines (Bamber, 1916; Watt, 1972; Conard, 1905). Chung & Ripperton (1929) regarded it as a unique vegetable of the orient. They have further stated that large quantities of the rhizomes were shipped to the United States to feed the oriental population. A soup is also prepared from the rhizomes of *Nelumbo* (Irvine & Trickett, 1953). Another type of soup is prepared with lotus seeds and red beans. It is served at banquets for blessing newlyweds to have a child the very next year. The soup is also presented at New Year festivals in China.

Rhizomes are bought wholesale in China, boiled and sold in the form of slices or dried for future use and for transportation elsewhere. The native American tribals in North America roast the seeds of *Nelumbo lutea* and extract oil from them by crushing, boiling and cooling and then skimming off the oil. The seeds are also ground into meal for soup or mush. The rhizomes are used in the preparation of a kind of arrowroot which is aromatic, sweet in taste, nutritious and a tonic too.

In India, two types of rhizomes: white and reddish brown are generally available in the market. They are sold as vegetables and commonly called 'Kamal kakri' or 'Bhen'. Rhizomes are eaten raw or after roasting while dried slices of rhizome are used in curry or fried as chips. Rhizomes are also pickled in slices which tastes delicious. They may be frozen and used as an ingredient in pre-cooked foods (Goel, 1999).

The rhizomes, petioles, leaves and flowers are used in savoury dishes. Seeds can be candied as sweet snacks. The thick, starchy rhizomes are boiled, stir-fried and pickled. Petals are often placed to garnish soups. Most delicious is a dish of 'Kamal kakri' (rhizome) cooked with 'Besan' (Gram powder). In Kashmir, lotus rhizomes are also cooked with meat and leafy green vegetables. Seeds are a popular ingredient in local desserts such as 'Cheng teng' which is very popular in Singapore. Lotus

White rhizome of Lotus

Reddish brown rhizome of Lotus

Green Fruiting (Torus) with a bud and a flower of Lotus

rhizomes are considered as a versitle food and used in cuisine as a delicacy. It is a powerhouse of nutrients like vitamins and minerals. Lotus rhizomes may help in managing blood pressure, reducing risk of heart attack and incuring various allergies and infections. In china lotus leaves are cooked as vegetable.

Green fruiting torus (*Kamal gatta*) containing carpels is a popular edible item in rural and suburban areas particularly in northern India. They are often sold fresh for the edible carpels embedded in it. Seeds are eaten raw, roasted, boiled or ground into flour after removing the intensely bitter seed coat and plumule. The carpels are a good source of cereal and of nutritive value.

The tender leaves, petioles and flowers are also eaten as vegetables. Highly valued perfume is extracted from lotus flowers. All parts of the lotus flowers and leaves of the *Nelumbo nucifera* are a traditional inebriating smoke, similar to mild cannabis. They are brewed into a delightful tea and are also steeped in wine. The stamens of lotus can be dried and made into refreshing herbal tea. The honey from bees which visit lotus flowers is reported to possess tonic properties and considered most useful for the cure of eye infections. Dry seed pods are in great demand in the dehydrated floral market for making various types of floral arrangements.

Table-3: Analysis of the matured dried carpels showing the following nutrient contents

Constituents	Availability (%)	Constituents	Availability (mg/100g)
Total Carbohydrate	66.6	Calcium	136.0
Protein	17.2	Phosphorus	294.0
Water	10.0	Iron	2.3
Sucrose	4.1		
Ash	3.8		
Fibre	2.6		
Reducing Sugar	2.4		
Fat	2.4		

(Source: Wealth of India, CSIR, 1966)

Table-4: Analysis of lotus rhizomes showing the following nutrient contents

Constituents	Availability (%)	Constituents	Availability (mg/100g)
Water	83.0	Thiamine	0.22
Starch	9.25	Riboflavin	0.06
Protein	2.70	Niacin	2.10
Sucrose	0.41	Ascorbic acid	15.00
Reducing Sugar	1.56		
Ash	1.10		
Fibre	0.80		
Fat	0.11		
Calcium	0.06		

(Source: Wealth of India, CSIR, 1966)

Lotus rhizomes for vegetable purpose

Torus with edible green seeds

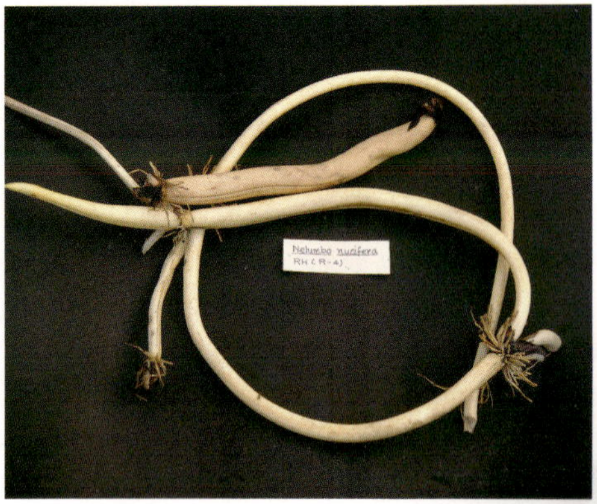

Young rhizome of *Nelumbo nucifera*

Lotus silk

Lotus silk is considered as one of the rarest and the most luxurious and beautiful fabrics in the world, produced only in small scale by the indigenous people all across Cambodia, Myanmar, and more recently in Vietnam since 2017. This natural fibre is only extracted by a few skilled ethnic craftsmen in these countries. But making lotus silk is not at all easy. Extracting enough lotus silk even for one scarf can take up to two months, and the final product can cost 10 times more as much as the regular silk.

Lotus silk is obtained from the matured leaf petioles and the floral pedicels which are selected and handpicked regularly by these locals from ponds and lakes. Each petiole/pedicel contains a minuscule amount of thin, sticky fibre, which must be rolled together. The threads need to be processed within 24 hours while they are still wet, otherwise, they will break. And so harvesting has to be done every day. The lotus crop is only available for harvest during April - September. Extracting the lotus fibre has been a very strenuous and troublesome task. The fibre is incredibly delicate too once it is dried, The threads are carefully weighed down and delicately hand-spooled. Then they are put into the loom immediately. These fibres are fragile, but once woven, can be as durable as the traditional silk or more shiny and elastic than the silk.

A 25 centimetres scarf can cost as much as 200 US $. A team of roughly 20 workers can create lotus fibres each day, allowing them to produce 10 to 20 scarves every month. This soft silk is breathable like linen, and slightly more elastic. Such luxurious traits have made it popular with tourists searching for the rare souvenirs. It has also been recently been picked up by a large and famous international fashion brands searching for new luxury fibre. Its production scale has been extremely limited because of only few trained ethnic people are involved in making of the lotus silk threads. Despite the hard work involved in producing lotus silk, indigenous people in these

Lotus leaf petiole and pedicel

Lotus silk thread harvesting from petiols and pedicels

Lotus silk thread

Lotus silk scarf

Lotus silk weaving

Lotus silk scarf

countries hope that their unique skill will one day grow into a larger and more lucrative industry. Our country can also take up this venture as a small scale industry and generate various job opportunities for the rural and urban masses as well as their economic empowerment.

LIST OF LOTUS VARIETIES

CHAPTER-25

Name	Size/Colour
'Alba Grandiflora'	White
'Angel Wings'	White
'Antou Hong'	Small, semi- double, red
'Antouchun'	Semi-double, red bowl
'Asiatic-album'	Single, red
'Ba Yi Lian'	Single, pink
'Baby Doll'	Dwarf, white
'Bai He'	Double, red tipped
'Bai Tiane'	Single, white
'Bai Wantian'	Single, white
'Bai Xianzi'	Semi-double, white
'Baichuantai'	Single, white
'Baihailian'	Double, white
'Baijianlian'	Single, white
'Baijunzi Lian'	Single, white
'Baimei Lian'	Small, double, pink
'Baimudan'	White
'Baiqianye'	Double, white
'Baishaoyaolian'	Double, white
'Baiwanwan'	Double, white
'Baixianglian'	Single, white
'Baixue gongzhu'	Double, white bowl
'Baiyi Zhanshi'	Single, white

Flowers of various Lotus cultivars

Name	Size/Colour
'Baiyun Wanlian'	Semi-double, white bowl
'Bawang Pao'	Small, double, versicolour
'Bayi Lian'	Single, pink
'Beauty'	Pink
'Beijing Wanlian'	Single, red tipped
'Bellagio'	Double, red
'Ben Gibson'	Pink/yellow base
'Benikunshiren'	Single, red
'Benimaihiren'	Single, pink
'Benyue'	Single, white
'Bilian'	Double, white

Nelumbo nucifera **(pink double) cv. 'Krishna' at CSIR-NBRI**

Name	Size/Colour
'Biluo liuhong'	Double, red
'Bing Xin'	Small, semi-double, white
'Bingijao'	Double, yellow
'Bingxin Yujie'	Small, double, white
'Bixue Danxin'	Single, red
'Bua Kem Champoo'	Single, medium pink
'Byakkohren'	Single, white
'Chutian Xiangyun'	Single, white

Name	Size/Colour
'Chuunichiyuugiren'	Single, red
'Culweixizhao'	Double, pink
'Caidie'	Semi-double, yellow red
'Carolina Queen'	Pink/yellow base
'Caspium'	Large, dark rose
'Chaitoufeng'	Double, pink
'Chang'e Ben Yue'	Small, double white
'Changban Fenlian'	Single, pink

Name	Size/Colour
'Changshou Xuezhe'	Single, white
'Charles Thomas'	Dwarf, medium pink
'Chawan Basu'	White, edged pink
'Chengxian Bilian'	Double, white
'Chengxian Fenlian'	Double, pink
'Chongban bayi lian'	Double, pink bowl
'Chongshuihua'	White
'Chongtai Lian'	Double, pink
'Chu Xiao'	Small semi-double, versicolour
'Chu'e Xishui'	Semi-double, yellow
'Chun Wang'	Small, semi- double, pink
'Chun Yan'	Small, double, white
'Chun Yu'	Small, semi- double, pink
'Chundulao'	Double, red
'Chunshan Fucui'	Semi-double, pink
'Chunshui Lubo'	Single, white
'Chushui Furong'	Double, pink
'Chunshui Huangli'	Semi-double, yellow bowl
'Da Bilian'	Double, white
'Da Feng'	Small, double pink
'Da Jin'	Double, red tipped
'Da Zilian'	Single, red
'Da Ziyu'	Small, double, red
'Dahongpao'	Single, red
'Dahou Lian'	Single, red
'Dajin'	Double, pink
'Dan He'	Small, semi-double, veriscolour
'Dan Niang'	Small, double, red
'Dan Yue'	Small, semi-double, white
'Dan Yun'	Small, double, pink
'Danfeng Chaoyang'	Small, single, versicolour

***Nelumbo nucifera* (pink double) cv. 'Krishna'**

Name	Size/Colour
'Dasajin'	Double, versicolour
'Debbie Gibson'	Cream
'Di Cui Lian'	Small, semi-double, white
'Dian e zhuang'	Single, white
'Dianjang Chun'	Double, pink
'Die Lian Hua'	Small semi-double, versicolour
'Donghu Bailian'	Single, white
'Donghu Chun Xiao'	Single, red
'Donghu Honglian'	Single, pink
'Donghu Xizhao'	Small, double, red
'Doulu Furong'	Double, white
'Du Li'	Single, white
'Echeng Hong Lian'	Single, red
'Emerald Daybreak'	Double, white with pink edges

***Nelumbo nucifera* (off white) double lotus**

Name	Size/Colour
'Empress'	White edged crimson
'Fei Hong'	Double, red
'Feiju'	Semi-double, yellow
'Feiyunqianye'	Double, pink
'Fen Chuan Tai'	Single, pink
'Fen Linglong'	Double, pink
'Fen Xia'	Double, pink
'Fenban Lian'	Small, single, pink
'Fenceng Tuo Cui'	Small, double, pink
'Fenchuantai'	Single, pink
'Feng Cai'	Small, semi-double, versicolour
'Feng Wu'	Small, single, versicolour
'Fenging Lian'	Single, white

Name	Size/Colour
'Fenhonglian'	Single, pink
'Fenlinglong'	Double, pink bowl
'Fenlou Chun'	Small, double, pink
'Fenloutai'	Double, pink
'Fenmian Taohua'	Small, double, red
'Fenqianye'	Double, pink
'Fensong qiu'	Double, pink bowl
'Fenwalian'	Single, pink bowl
'Fenwanlian'	Semi-double, pink bowl
'Fenxia'	Double, pink bowl
'Flavescens'	Yellow
'Fojianxiao'	Single, pink
'Foshou Guanying'	Small, single, versicolour
'Foshoulian'	Single, pink
'Fozuolian'	Double, pink
'Furong Qiuse'	Small, double, red
'Gan Bai Lian'	Single, white
'Gan Fu Da Hong'	Single, red
'Gang Hua'	Small, semi-double, red
'Genshibasu'	Single, red
'Giant Sunburst'	Large, yellow
'Glen Gibson'	Pink/yellow-orange base
'Gregg Gibson'	Lavender pink
'Guangchang Bai Ye'	Double, red
'Guangchang Lian'	Double, red
'Gudai Lian'	Single, pink
'Gui Fei'	Small, double, pink
'Gulang Xiaohong'	Semi-double, red bowl
'Gyokushuuren'	Double, dark red
'Gyozankohren'	Semi-double, deep red
'Han Xiao Lian'	Small, semi-double, red

Name	Size/Colour
'Hanjiang Dudiao'	Double, pink
'Hekidairen'	Double, white
'Himebasu'	Single, deep red
'Hong Chun'	Double, orange yellow with purple apices
'Hong Lou'	Double, red
'Hong Mudan'	Small, double, red
'Hong Ri'	Single, white
'Hong Shaoyao'	Double, red
'Hong Wanlian'	Double, red bowl
'Hong Xiuqiu'	Semi-double, red tipped
'Hongbian Yudie'	Single, white
'Honghu Honglian'	Single, red
'Honglian Lian'	Single, red
'Honglingjin'	Single, red
'Hongqianye'	Double, red
'Hongwalian'	Semi-double, red bowl
'Hongwanwan'	Double, red
'Hongxia'	Single, red
'Hongxianglian'	Single, pink
'Hongyan Ningxiang'	Semi-double, red
'Hongzhan tuozhu'	Double pink bowl
'Huadeng Chushang'	Semi-double, red bowl
'Huang Li'	Small, semi-double, yellow
'Huang Ying'	Small, semi-double, white
'Huangwufei'	Single, yellow
'Huo Ju'	Small, double, red
'Huohua'	Single, red bowl
'Ittenshikai'	Single, white with pink stripes
'Jianwu'	Single, yellow
'Jianxiang Yunu'	Single, white

***Nelumbo nucifera* (dark pink) flower**

Name	Size/Colour
'Jiao Rong'	Small double, pink
'Jiaorong Sanbian'	Double, pink
'Jiaorong Wanlian'	Single, pink bowl
'Jiaorong Zuibei'	Double, pink bowl
'Jiefang Hong Lian'	Single, red
'Jifeilian'	Single, red
'Jin Xiu'	Small, double, pink
'Jinbianlian'	Double, pink bowl
'Jinfeng Chuchao'	Single, yellow
'Jinfeng Zhanchi'	Single, yellow
'Jinzhou Lian'	Small semi-double, white
'Jinzhu Luo Yupan'	Small, single, white
'Ju Lian'	Single, pink

Nelumbo nucifera **(single white)**

Name	Size/Colour
'Jubo'	Double, pink bowl
'Kashohren'	Single, white with red tip
'Kermesinum'	Bright pink
'Kinzuiren'	Double, red
'Kohchohdairen'	Double, red
'Kohmanman'	Double, red
'Konrinren'	Single, white with red tip
'Krishna'	Double, pink
'Kuihua Xiangyang'	Semi-double, pink bowl
'Lang Hua'	Small, single pink
'Lavender Lady'	Large, lavender
'Lian Xia'	Small, semi-double, Red
'Linda'	Deep pink flushed w/yellow

Name	Size/Colour
'Longfei'	Semi-double, white
'Longhua Lian'	Small, semi-double, yellow
'Longzhu'	Double, red
'Lu Ban Chun'	Small, double, pink
'Luhuanong'	Double, pink bowl
'Luo Shen'	Small, single, pink
'Luoxia Yingxue'	Double, pink
'Lushan Bailian'	Semi-double, red tipped
'Lutea Yellow Bird'	Single, yellow
'Ma Nao'	Small, double, versicolour
'Maggie Belle Slocum'	Mauve
'Maihiren'	Single, yellow with red tip
'Maiyohren'	Single, red
'Makotobasu'	Double, red
'Manaohong'	Single, red
'Manjianghong'	Double, red bowl
'Mario Zanfardino'	Single, cream
'Meikohren'	Single, pink
'Meiyuan Xiu Se'	Small, semi-double red
'Mei-zhong hong'	Single,versicolour
'Meizhou Huanglian'	Single, yellow
'Mohe'	Single, red
'Moli Lian'	Small, single, yellow
'Momo Botan'	Dwarf, double red
'Mrs. Perry D. Slocum'	Double, red changing to creamy yellow
'Mudan Xianzi'	Double, white bowl
'Nihelu Lian'	Single, pink
'Nihon Kodaibasu'	Single, red
'Nikki Gibson'	White yellow base pink tips
'Nishang Xiaoqu'	Double, pink bowl
'Nishangqu'	Single, pink

Name	Size/Colour
'Ohjibasu'	Single, yellow
'Ohren'	Single, yellow with red tip
'Osiris'	Deep rose
'Patricia Garrett'	Pink
'Pekinensis Rubra'	Bright rose carmine
'Perry's Giant Sunburst'	Large, light yellow
'Perry's Super Star'	Double, pink changing to creamy yellow
'Pingshan Furong'	Double, pink
'Pink Radiance'	Pink bowl
'Poontarik'	Single, white
'Pulchra'	Single, pink
'Pygmaea Alba'	Dwarf, white
'Qi Xia Nu'	Small, semi-double, pink
'Qian Bai Wan'	Double, white
'Qian Dui Jin'	Small, double, red
'Qianbaiwan'	Double, white
'Qianban Lian'	Large, rosy red, thousand petals
'Qianling Baihe'	Double, white
'Qilu Furong'	Semi-double, pink
'Qing fang Lian'	Small, semi-double, yellow
'Qinglin Honglian'	Single, red
'Qintai geshou'	Double, pink bowl
'Qiu Cui'	Small, semi-double, yellow
'Qiushui Changtian'	Double, pink
'Remnian Taohua'	Small, double versicolour
'Rin'nohren'	Single, white
'Rongjiao'	Double, pink
'Rosea'	Deep rose pink
'Rosea Plena'	Double, bright rose

Nelumbo nucifera **(white double)**

Name	Size/Colour
'Rozanpakuren'	Double, white
'Rubra'	Red
'Ruyan Huan'	Small, double, pink
'Ruyi Lian'	Single, pink
'Sai Jin'	Small, double, versicolour
'Sai Meigui'	Single, pink
'Saienjishoren'	Double, white
'Saifuozuo'	Double, pink
'Saikohjibyakuren'	Single, white
'Saikoren'	Double, white
'Seigetsuren'	Single, white
'Sengmao Lian'	Small, semi-double pink
'Shancha Hong'	Double, red bowl

***Nelumbo nucifera* cv. 'Momo-Botan'**

Name	Size/Colour
'Shanglin Chunse'	Small, semi-double, pink
'Shaoxing Honglian'	Single, pink
'Sharon'	Double, pink
'Shengxian Fenlian'	Double, pink

Name	Size/Colour
'Shin'nyoren'	Single, white
'Shirokunshishoren'	Single, white
'Shiroman'	Double, cream
'Shouxingtao'	Double, pink
'Shu Honglian'	Single, red
'Shuangchen Yue'	Small, double, pink
'Shuang'e Diecui'	Semi-double, white
'Sokuhiren'	Single, red
'Strawberry Blonde'	Pink to yellow
'Su Di Chun Xiao'	Small, semi-double, versicolour
'Suihiren'	Single, white with red tip
'Sunwen Lian'	Single, red
'Suzanne'	Medium pink
'Taibai Lian'	Single, white
'Taihakuren'	Single, white
'Tangwan'	Semi-double, pink
'Tangzhaodisi lian'	Double, red
'Tao Hong So Yu'	Small, double, red
'Tenghu Lian'	Single, pink
'Tenjikumadarabasu'	Single, white with red stripes
'Tenjohren'	Single, red
'The Queen'	Cream
'Tiangao Yundan'	Double, white
'Tianjiao'	Double, pink
'Tiannu Sanhua'	Single, yellow red
'Tiansan Bitai'	Double, white
'Tohkoren'	Single, red
'Tohren'	Single, red
'Tong Yan'	Small, double, pink
'Tongxiumian'	Single, pink bowl
'Tulip'	White, tulip shaped

Name	Size/Colour
'Vietnam Red'	Red
'Wan-er Hong'	White
'Wanshou Hong'	Double, red
'Waren'	Single, red
'Wawalian'	Single, white bowl
'Wawalon'	Dwarf, single white
'Wei Yu'	Small, double, white
'Weishan Honglian'	Single, red
'Welcoming'	Pink
'Wucai Feng'	Double, red
'Wufeilian'	Single, versicolour
'Xi Hongyi'	Small, double, pink
'Xi Ying Men'	Small, semi-double, red
'Xiaguang ranzhi'	Single, pink bowl
'Xiamen Wanlian'	Single, white bowl
'Xiang Jian Huan'	Small single pink
'Xiang Yu'	Double, red
'Xiangshan Lian'	Small, single white
'Xiangshan Wanlian'	Semi-double, white
'Xiangsi Dou'	Small, double, red
'Xianjiao'	Double, pink
'Xiannu Sanhua'	Single, pink
'Xiao Foshou'	Small, single red
'Xiao Jin'	Double, red tipped
'Xiao Li'	Small double, red
Xiao Liren	Small, double, pink
'Xiao Mei'	Small, semi-double, pink
'Xiao Mei Hong'	Small, single, versicolour
'Xiao Sanse Lian'	Semi-double, yellow-red
'Xiao Tianshi'	Small, semi-double, pink
'Xiao Ying'	Small, semi-double, white

Name	Size/Colour
'Xiao Zhizi'	Small, semi-double, white
'Xiaobilian'	Double, white
'Xiaobitai'	Double, white bowl
'Xiaofeng liangyue'	Double, pink
'Xiaojinfeng'	Single, yellow bowl
'Xiaolianzuo'	Double, pink bowl
'Xiaoqiao'	Double, pink
'Xiaoshouxing'	Single, red bowl
'Xiaowufei'	Single, versicolour
'Xiaoxia'	Double, red
'Xiaoye'	Double, pink
'Xiaoyulou'	Double, white bowl
'Xiaozuixian'	Double, pink bowl
'Xidan Hong'	Single, red
'Xiezhua'	Small, semi-double, white
'Xihu Honglian'	Single, red
'Xingguang'	Semi-double, white bowl
'Xinghua Chunyu'	Double, pink
'Xinhua Nu Fang'	Small, double, red
'Xinlingmei'	Double, pink bowl
'Xishang' meishao	Single, pink bowl
'Xiwang'	Small, semi-double, white
'Xixiangfeng'	Single, red
'Xiyang Hong'	Small, double, versicolour
'Xuanwu Honglian'	Single, red
'Xuewu'	Semi-double, white bowl
'Yaechawanbasu'	Double, red
'Yan er Mei'	Single, white
'Yan Li'	Small, double, pink
'Yan Xia'	Small, double, red
'Yangzhou Bowl'	Pink

Name	Size/Colour
'Yanwu Yingdi'	Small, double, versicolour
'Yanxinzhuang'	Double, pink
'Yanyangtian'	Single, red
'Yao Hua'	Small double, versicolour
'Yeming Zhu'	Single, white
'Yi Ban Chun'	Small, semi-double, pink
'Yi Peng Xue'	Small, semi-double, white
'Yidai Jiankuan'	Double, pink
'Yidianhong'	Single, red
'Ying Tao Wan Lian'	Small, semi-double, pink
'Ying Ying'	Small, semi-double, yellow
'Yingbin Furong'	Single, white bowl
'Yingerhong'	Semi-double, pink bowl
'Yingguang'	Small, semi-double, white
'Yinhongqianye'	Double, pink
'Yizhangqing'	Single, white
'Youyhong'	Dwarf, single pink
'Youyihong erhao'	Small, double, pink bowl
'Youyihong sanhao'	Single, pink bowl
'Yu Foshou'	Small, single, versicolour
'Yu Hua'	Small, double, white
'Yu Li'	Small, double, versicolour
'Yuanyan Jin'	Small, single, versicolour
'Yulou renzui'	Double, pink bowl
'Yuloutai'	Double, white bowl

Name	Size/Colour
'Yun Teng Xia Wei'	Small double, pink
'Yunzhong Feng'	Double, pink
'Yuwan'	Double, white bowl
'Yuxiu Lian'	Double, red
'Zepan Furong'	Semi-double, pink
'Zhao Yun'	Small, double, pink
'Zhong-ri youyi lian'	Single, red
'Zhu Fu'	Small, double, red
'Zhu Guang'	Small, semi-double, pink
'Zhulian Ban Juan'	Small, semi-double, pink
'ZhuoanYing Hui'	Small, double, red
'Zhuoshanglian'	Double, red bowl
'Zhuyi Shizhe'	Small, semi-double, red
'Zhuzhou Honglian'	Single, red
'Ziyou Zhong'	Small, semi-double, pink
'Ziyulian'	Dwarf, double, pink purple
'Zui Dongfeng'	Small, double, pink
'Zui Li Hua'	Small, semi-double,versicolour
'Zui Wu'	Single, white
'Zuibanxun'	Single, pink
'Zuibei'	Single, yellow-red
'Zuikohren'	Single, white with red tip
'Zuitun'	Double, pink
'Zuiyun'	Double, pink

Source: (Kit & Ben Knotts, Cocoa Beach Florida)

LOTUS FOR BIOREMEDIATION

CHAPTER-26

Lotus has the ability to uptake heavy metals and can be grown in ponds receiving industrial effluents to serve the purpose of water purification in aquatic bodies. The lotus has high tolerance for sulphur dioxide and can be planted in ponds around the industrial regions for bioremediation.

Chromium is one of the most toxic metals widely found in water bodies. The large scale use of chromium in metallurgical pigment and dye, textiles and electroplating, makes these industries potential sources of chromium pollution. According to a recent estimate *ca.* 2000 to 3200 tonnes of chromium escapes into the Indian environment annually from the tanning industries alone (Thiagrajan, 1992). The chromium level in effluents usually ranges between 2000 to 5000 µg ml/l against the recommended permissible limit of 2.0 µg ml/l. Contamination of water resources by the effluents is posing a threat to the aquatic ecosystem. Chromium interferes with several metabolic processes of plants and animals and is highly toxic to plants resulting in the reduced roots, phytomass and photosynthetic pigments, chlorosis, stunting and eventually plant death. Aquatic plants have shown greater potential for ameliorating the metal load of waste water by active uptake and surface adsorption. Rooted macrophytes with floating foliage may have the key role in cleaning aquatic ecosystems as their soil binding roots, rhizomes and stolons help to reduce erosion and facilitate colonies of benthic algae and invertebrates. These plants are also useful in abating and monitoring of toxic metals as they do not migrate and attain equilibrium with their surroundings very quickly.

Nelumbo nucifera Gaertn. a rooted macrophyte, has been selected for the present study because it is known to grow luxuriantly in chromium polluted water bodies, eutrophic lakes and most importantly because it occurs throughout India even up to 1800 m altitude. It has been considered pertinent to evaluate metal uptake potential and related toxicity in the plant before exploiting the species in a phytoremediation system for amelioration of metals from effluents or polluted water bodies.

Lotus aquatic body in CSIR-NBRI Botanic Garden, Lucknow

N. nucifera was able to accumulate about 79% chromium after 168 hours treatment from 50 μm chromium concentration. Rooted aquatic plants adapt themselves to a heavy metal contaminated environment. Although rooted aquatic plants with floating foliage are important for cleaning lentic ecosystem, they should be exploited for the biological treatment of polluted water. Results have shown that rooted macrophytes with floating foliage are well suited to phytoremediation of heavy metal contaminated water. *N. nucifera* showed a high potential of chromium amelioration from the waste water by accumulating it in varying amounts (roots>leaves>rhizomes). Results indicate that the plants of *N. nucifera* have high chromium uptake potential and the species could be exploited for removing chromium from polluted water bodies. Nitrate reductase activity and total chlorophyll content in the plants may be used as an early indicator of chromium injury in the aquatic plants (Vajpayee *et al.*, 1999).

Nutrients, especially nitrogen and phosphorus, have been recognised as the pollutants that contribute to algal bloom as eutrophication in lakes and slow moving water. A pilot scale aquatic pond, 1.8 m in length, 0.6 m in width and 1.2 m in depth, was constructed at the Hat Yai Municipality Central Wastewater Treatment Plant in Thailand. For the removal of pollutants from the domestic wastewater, *Nelumbo nucifera* and *Hydrilla verticillata* were planted together in the ponds. These plant species were found effective in the pilot scale experiment. The aquatic plant system has been counted as one of the processes for wastewater recovery and recycling. Low cost and easy maintenance make the aquatic plant system attractive to use. Thus, constructed ponds with aquatic plants are increasingly applied as a viable treatment for municipal wastewater (Kanabkaew & Puetpaiboon, 2004).

Fast degrading aquatic bodies near urban localities due to pollution

Mass Flowering of Lotus in an aquatic body

LOTUS IN LANDSCAPING

CHAPTER-27

The lotus growing in pond or in a pot gives great satisfaction. A pond full of lotus flowers is certainly a captivating sight. Due to environmental degradation, wetlands are decreasing so the lotus population is also threatened. Lotus can be grown in formally or informally designed ponds in botanic gardens, parks, picnic spots and industrial regions. Lakes and ponds planted with lotus varieties will give a boost to eco-tourism. In these days urban life is full of noise and pollution; people want a fresh and clean environment. India is a vast country and there is great scope for the development of eco-tourism. Lotus and fish culture is a good combination for balancing the eco-system and improving the economic condition of the local people as well as progressive farmers. Lotus growing in lakes and ponds also attracts cranes, ducks and other birds as there is good food for their survival. Restaurants and motels should be established near lotus lakes to cater for the needs of tourists by providing them with various lotus dish preparations.

Botanic gardens in India should enrich indigenous *Nelumbo* races/cultivars from within the country. There are so many colourful and attractive cultivars available in Thailand, Japan, China and USA, which can also be introduced in parks and gardens. Lotus blooms from July to September when there is a dearth of flowers in this season. Lotus flowers have a mild fragrance that is soothing and gives mental peace and pleasure to a passerby. Lotus blooms also help in providing pleasing and healthy environment. It gives us strength to lead a clean life in the present day miry world.

A view of lotus blooming in various chambers of aquatic body at CSIR-NBRI Botanic Garden

For the proper conservation of the lotus, theme gardens should be developed aesthetically with tabloids providing information on all aspects of the plant. There should be chapters in environmental education books for students to learn about the multiple uses and importance of our 'National Flower.'

For promoting the lotus culture exhibitions should be organized in parks and gardens in the rainy season when lotus ponds are in full bloom. Organization of flower shows and exhibitions blends science and art. These are good opportunities where the garden lovers and a cross section of society can interact and appreciate the beauty of the lotus, the flower of paradise. Exhibitions can also have sections for lotus photographs, painting and handicrafts. For promoting the food value of the lotus there should also be restaurants at the show ground where delicious and nutritious lotus preparations can be served to the visitors.

Lotus Growing in Pots

Lotus may be grown in earthen pots, which should be 60 cm high and 60 cm in diameter. Dwarf lotus varieties can be grown in pots of 30 cm high and 30 cm in diameter. Lotus plants can also be grown in plastic containers. A 20-25 cm thick layer of 2:1 clay soil and farmyard manure mixture is placed at the bottom of the pot. A rhizome segment is placed horizontally at a slight angle and is buried 5.0 cm deep in the soil. The growing tip of the rhizome should be left exposed. Water is added to the pot, which should not initially be more than 10 cm above the soil mixture. The water level is gradually raised keeping 0.5 cm below the rim of pot. Pots should be kept in an open place because the lotus plants prefer full sun shine. Lower temperatures may cause the plants to go into dormancy and it will be difficult to get blooms. 'Baby Doll' and 'Momo Botan' are ideal dwarf lotus cultivars for growing in pots and tubs. To control mosquito larvae a few goldfish may be dropped in the container. This also adds additional attraction to the lotus container / pond.

Lotus growing in various chambers of a aquatic body at CSIR-NBRI Botanic Garden, Lucknow

Lotus growing in a pot

LOTUS: THE CROP OF FUTURE

CHAPTER-28

There is heavy demand of lotus flowers in July-September for woripping Lord Shiva. Lotus flowers near Shiva temples are sold at Rs. 5/- 10/- per unit. Considering its ornamental, medicinal and economic importance, lotus is the crop of the future. Low lying areas in fields should be used for lotus cultivation. To popularise the lotus crop, Horticulture Departments and Agricultural Universities should take up research and developmental programmes. There should be model layouts giving details of crop cultivation as well as techno-economics. Progressive farmers should be invited to participate in seminars and field trips should be conducted for the entrepreneurs. They should be provided with literature on the crop. Post-harvest technology for cut lotus flowers and rhizomes is a very important factor as lotus is deeply associated with our culture. There is a heavy demand for lotus flowers in temples and for marriage ceremonies. The dry seed pods fetch good prices as excellent dehydrated items in floral decorations.

Rhizomes, young petioles, leaves and fresh seeds comprise the edible items. The significance of their food value should be imparted especially to the urbanites. Co-operative societies could also be involved in popularizing and marketing the lotus

Lotus decoration in a temple

Lotus flower decoration in a marriage venue

crop. Yun, *et al.* (2011) found immunoregulatory and anti-HIV-1 enzyme activities of antioxidant components from the lotus rhizomes.

With proper management, the yield of rhizomes can be raised to 10 quintals per hectare. The lotus crop has enormous potential for commercial exploitation and profit earning. From the sale of rhizomes, flowers, carpels and seeds one can earn a net profit of Rs. 50,000/- per hectare. Apart from the domestic market, there is good scope for exporting rhizomes to the South-East Asian countries.

LOTUS –IMMORTALIZED ON STAMPS AND COINS

CHAPTER-29

For creating awareness and concern towards conservation of nature and natural resources amongst the masses, one of the best and emphatic mediums has been engraving and printing images of plants/animals on coins and postal stamps. Releasing coins, currency notes as well as postal stamps with pictures of a certain plant or animal species attributes special emphasis of a country about its deep concern for the conservation of its natural resources. Several countries have released beautiful postal stamps as well as currency notes/ coins with photographs of significant species of economic, medicinal and rare plants and animals to commemorate various occasions. Countries like India, Singapore and South Africa need special mention in this direction for giving a special message about their sincere commitment towards conservation of their precious natural resources. Canada, China, Japan and Vietnam have also released the postal stamps and coins with photos of some important plant/ animal taxa during special events, thereby displaying their deep concern for conservation of nature. It is pertinent that the Highest Gallantry award in the country in face of enemy, PVC (Param Vir Chakra) bears two Lotus flowers on its reverse.

Lotus Symbolized in Stamps

Lotus, (*Nelumbo nucifera*) has been embodied on postal stamps of India, Japan, Canada, China and Vietnam in South-East Asian region. On Indian postal stamps, lotus has appeared significantly on several occasions attributing importance to this plant species in culture as well as religion. India's National Flower, lotus (*Nelumbo nucifera*) was depicted explicitly on a Canada Post stamp, miniature sheet and stamps coil issued in March 2018. The beautiful stamps marked spring's return focusing on two varieties of Lotus including the Sacred Indian Lotus (*Nelumbo nucifera*), with delicate pink and white petals and of deep religious significance to Hindus and Buddhists.

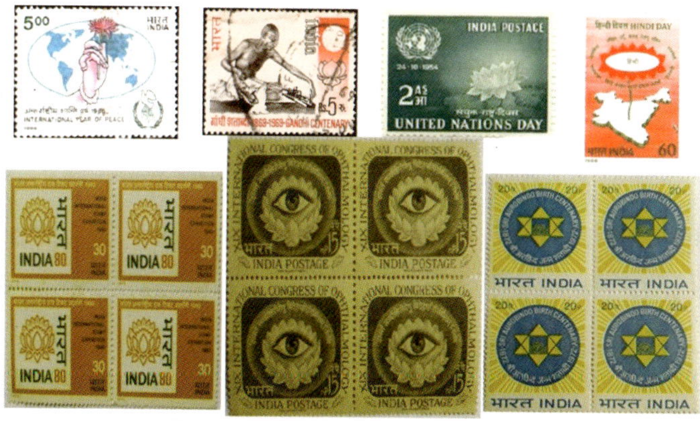

Lotus flower on postal stamps

Lotus pictures on postal stamps

India, a country hailed as 'Sujalam Suphalam' in its National song, is a beautiful superposition of flora, fauna and culturally diverse population! Having at least three major biodiversity hotspots, India has a variety of geographical zones which produce unique specimens of flowering plants. In an attempt to celebrate this floral variety, India Post had issued a set of 4 commemorative postage stamps on 1st July, 1977 including a 25 paise stamp depicting our National Flower, Lotus *Nelumbo nucifera*, Indian Lotus, Kamal or Padma is found growing in stationary water bodies such as ponds and lakes etc. It is of immense significance for the conservation of water resources on this planet.

Lotus Symbolized on Coins

A number of countries have applauded the Lotus species/ varieties/races by issuing Commemorative Stamps on various occasions in synchronism with multifarious properties associated with this extra ordinary flower. Coins / Currency Notes have been beautifully depicted with lotus flowers.A one rupee silver coin released during pre-independence era in Indiain 1917 by the then government, has a lotus flower on top on one side and their Emperor on the other side. All these efforts provide a clear message about the significance of lotus and its immense potential for further bio-prospection in future. Lotus (*Nelumbo nucifera* and *N. lutea*) still have a lot of scope to harness its economic and medicinal potential in the form of Lotus silk as well as smart cloths with Lotus effect which are the need of the future and will empower the economies of many nations. In China, coins symbolize seasons with lotus representing the summers.

Lotus figures on various Indian coins

REFERENCES

Agarwal, M., Dixit, V.P. and Sandhu, J.S. (1980). Possible mechanism of antifertility activity of 3-chloro-1, 2-Propanediol (U-5897) on the female genital tract of Rattus rattus Rufescens-A Biochemical and Histophysiological study, Proc. Indian Nat. Sci. Acad, 46 : 293-301.

Agnihotri, V.K., ElSohly, H.N. and Khan, S.I. (2008). Constituents of *Nelumbo nucifera* leaves and their anti-malarial and antifungal activity, Phytochem Lett, 89-93.

Angiosperm, Phylogeny Group (2016). An update of the Angiosperm Phylogeny Group Classification for the orders and families of flowering plants, APG IV, Botanical Jour. of the Linnean Soc, 181 (1): 1-20.

Anonymous, (1966). The Wealth of India: Raw Materials 7, Pub. Inf. Dir., CSIR, New Delhi, 7-9.

Arber, A. (1920). Water plants: a study of aquatic angiosperms, University Press, Cambridge, 1.

Arora, R.K., Mehra, K.L. and Nayar, E.R. (1980). Taxonomy of Cultivated Plants-Problems and Prospects, in Modern Trends in Plant Taxonomy, Glimpses in Plant Research – V (ed. P.K.K. Nair), 296-305.

Arthur, H.R. and Cheving, H.T. (1959). J. Chem. Soc, 2306.

Babu, C.R. and Gopinathan, M.C. (1986). The scenario of plant genetic resources, Bull. Bot. Surv. India, 28 (1-4): 159-168.

Bailey, L.H. and Bailey, E.Z. (1976). *Nelumbo lutea* Pers, Hortus Third-A Concise Dictionary of Plants Cultivated in the United States and Canada, p. 1290, 774.

Barthlott, W. and Neinhuis, C. (1997). Purity of the sacred lotus or escape from contamination in biological surfaces, Planta, 202 : 1-8.

Barthlott, W., Neinhuis, C., Jetter, R., Bourauel, T. and Riederer, M. (1996). Waterlily, poppy or sycamore: on the systematic position of *Nelumbo*, Flora, 191 : 169-174.

Behera, K.K. (2006). Plants used for Gynecological Disorders by Tribals of Mayur Bhanj District, Orissa, India, Internet Web Journal, 1.

Berry, E.W. (1917). Geologic History indicated by the Fossiliferous Deposits of the Wilcox Group (Eocene) at Meridian, Mississippi, U.S. Geol. Surv. Proc. Paper, 108 (E): 61-72.

Biswas, K. and Calder, C.C. (1954). Handbook of Common water and marsh plants of India and Burma 2nd edition, Rep. 1984. Bishen Singh Mahendra Pal Singh, Dehradun.

Borsch, T. and Barthlott, W. (1994). Classification and distribution of the Genus *Nelumbo* Adans. (Nelumbonaceae), Beitr. Biol. Pflanzen, 68 : 421-450.

Boyd, C.D. (1968). Fresh water plants: A potential source of protein, Econ. Bot, 22 : 359-368.

Brickell, C.D., Voss, E.G., Kelly, A.F., Schneider, F. and Richens, R.H.(Eds.) (1980). International Code of Nomenclature for Cultivated Plants (ICNCP), Royal Botanic Gardens, Kew, UK, International Code of Nomenclature for Cultivated Plants (ICNCP), Royal Botanic Gardens, Kew, UK, 1.

Burkill, I.H. (1946). On the dispersal of the plants most intimate to Buddhism, J. Arn. Arb., 27 (4): 327-339.

Burkill, I.H. (1965). Chapters on the History of Botany in India, Govt. of India press, Delhi.

Caius, J.F. (1939). Medicinal and poisonous plants of India: Waterlilies, Poppyworts, Fumitories, J. Bombay Nat. Hist. Soc., 40 (3&4): 513-527.

Campbell, C.L., Huang, J.S. and Payne, G.A. (1980). Defence at the perimeter: the outer walls and the gates. In: Horsfall JG. Cowling EB, eds., Plant diseases an advanced treatise, London: Academic Press, 103-120.

Chen, G., Zhu, M. and Guo, M. (2019). Research advances in traditional and modern use of *Nelumbo nucifera*: phytochemicals, health promoting activities and beyond, Crit. Rev. Food Nutr., 59 (Supl 1): 189-209.

Chen, W.P. and Zhang, S.H. (1989). The Structure, elongation and thickening of rhizome of *Nelumbo nucifera* Gaertn., Acta Bot. Sin., 31 (3): 191-197.

Chopra, R.N., Chopra, I.C., Handa K.L. and Kapur, L.D. (1958). Chopra's Indigenous Drugs of India, p. 679, U.N. Dhur & Sons Pvt. Ltd., Calcutta, 1.

Chopra, R.N., Nayar, S.L. and Chopra, I.C. (1956). Glossary of Indian Medicinal Plants, p. 115 , PID, CSIR, New Delhi, 1.

Chung, H.L. and Ripperton, J.C. (1929). Utilization and composition of oriental vegetables in Hawaii, United States Department of Agricultural experiment station, Bulletin, 60 : 45-46.

Conard, H.S. (1905). The Waterlilies, p. 279, The Carnegie Institution, Washington, 1.

Conventry, B.O. (1927). Wild Flowers of Kashmir (Series-I), p. 100, Raithby, Lawrence & Co. Ltd., London, 1.

Dumortier, B.C. (1829). Analyse des Familles des Plantes, Tournay, 1.

Eller, B.M. (1977). Road dust induced increase of leaf temperature, Environmental Pollution , 13 : 99-107.

Everett, Thomas H. (1980). The New York Botanical Garden Illustrated Encyclopedia of Horticulture, Garland Pub. Inc., New York, USA, 1.

Gebrie, E., Makonnen, E., Debella. A., Zerihun, L. (2005). Phytochemical screening and pharmacological evaluations for the antifertility effect of the methanolic root extract of Rumex steudelii , J Ethnopharmacol, 96 : 139-143.

Goel, A.K. (2001). Lotus: God's favourite bloem (Dutch), Hindorama (Netherlands), 2 : 26-29.

Goel, A.K. (1999). Lotus: Food for the body, Hinduism Today (USA), 7 : 30-33.

Goel, A.K. (1999). Lotus: God's Favourite Flower, Hinduism Today (USA), 7 : 35.

Goel, A.K., Sharma, S.C. and Sharga, A.N. (2001) The conservation of the diversity of *Nelumbo* (Lotus) at the National Botanical Research Institute, Lucknow, India, Bot. Gard, Consv. News. (U.K.), 3 (6): 52-54.

Goel, A.K., Swarup, V. and Roy, R.K. (2003). Lotus (II), in Commercial Flowers (Eds. T.K. Bose, L.P. Yadav, P. Pal., V.A. Parthasarthy & P. Das), Noya Udyog, Kolkata, 2 : 419-428.

Goel, A.K. (2000). Ex-situ Conservation of Diversity of *Nelumbo nucifera* (Lotus), Jour. Non-Timb. For. Prod., 7 (3/4): 161-165.

Green, C. (2002). Jewels of the pond, Horticulture, 99 : 50-55.

Gruter, W., Burdet, H.M., Chalower, W.G., Demoulin,V., Grolle, R., Hawksworth, D.L., Nicolson, D.H., Silva, P.C., Stafleu, F.A., Voss, E.G. and Mc Neill, J. (1988). International Code of Botanical Nomenclature, Koeltz. Konigstein, 1.

Gupta, S.M. (1971). Plants myths and traditions in India, Netherlands Brill Publishers, 65-67.

Hall, T.F. and Penfound, W.T. (1944). The Biology of American Lotus, *Nelumbo lutea* (Willd.) Pers, American Hidland Naturalist, 31 (3): 744-758.

Harlan, J.R. and De Wet, J.M.J. (1971). Towards a rational classification of cultivated plants, Taxon, 20 (4): 509-517.

Heijny, S. (1978). Conservation of plant communities in fish pond littoralis, pp. 429-433, in Dykyjova et al. (eds.), Pond Littoral Ecosystems, Springer, New York, 429-433.

Heywood, V.H. (ed.) (1978). Flowering plants of the world, p. 335, Oxford University Press, London, 1.

Holloway, P.J. (1994). Plant cuticles: physicochemical characteristics and biosynthesis. In: Percy KE, Cape JN, Jagels R, Simpson CJ (eds), Air pollution and the leaf cuticle. Springer, Berlin, 1-13.

Hong, T.D., Linington, S. and Ellis, R.H. (1998). Compendium of Information on Seed Storage & Behaviour, RBG, Kew (UK), 2 : 593.

Hooker, J.D. (1872). The Flora of British India, Rev. & Co. Ltd., London, UK, 115-116.

Hsu, S.Y. (1945). Medicinal plants of Chengtu herb shops, J. West China Border Res., 15 (B): 95-177.

ICMR, (2017). Reviews on Indian medicinal plants, Na-Ny, *Nelumbo* Adans., Indian Council of Medical Research, New Delhi, 17 : 95-144.

Irvine, F.R. and Trickett, R.S. (1953). Waterlilies as food, Kew Bull., 3 : 363-370.

Ito, M. (1987). Phylogenetic systematics of the Nympheales, Bot. Mag, 100 : 17-35.

Jain, A., Singh, R.S. and Singh, B. (2004). Natural Product Radiance, 3 (6): 418-421.

Jain, S.K. (1990). Conservation of aquatic plants, Ecology and Management of Aquatic Vegetation in the Indian Subcontinent, (ed. Brij Gopal), 237-242.

Jeffrey, C.E. (1968). Systematic Categories for Cultivated Plants, Taxon, 17 : 109-114.

Jeffrey, C.E. (1986). The cuticle, epicuticular waxes and trichomes of plants, with reference to their structure, functions and evolution. In: Juniper B.E., Southwood S.R., eds, Insects and the plant surface. London: Edward Arnold, 26-63.

Jha, V.N., Kargupta, A.N., Dutta, R.N., Jha, U.N., Mishra, R.K. and Saraswati, K.C. (1991). Utilization and conservation of Euryale ferox Salisb. In Mithila (North Bihar), India, Aquatic Bot., 39 : 295-314.

Jones, J.A. (1928). Overcoming delayed germination of *Nelumbo lutea*, Bot. Gaz., 85 : 341-343.

Juniper, B.E. (1991). The leaf from the inside and the outside: a microbe's perspective. In: Andrews J.H., Hirano S.S., eds, Microbial ecology of leaves. New York: Springer, 21-42.

Kachroo, P. (1984). Aquatic Biology in India. Bishen Singh Mahendra Pal Singh, Dehradun.

Kak, A.M. (1985). Family Nympheaceae in the North Western Himalaya (Aquatic and wetland vegetation of North Western Himalaya-XXI), J. Econ. Tax. Bot., 7 (3): 591-598.

Kanabkaew, T. and Puetpaiboon, U. (2004). Aquatic plants for domestic wastewater treatment: Lotus (*Nelumbo nucifera*) and Hydrilla (*Hydrilla verticillata*) systems, Songklanakarin J. Sci. Technol., 26 (5): 749-756.

Kasumi, M. and Sakuma, F. (1998). Flowering, pollination. Fertilization and seed formation in lotus rhizome plant. (Japanese), Journal of the Japanese Society for Horticulture Sciences, 67 (4): 595-599.

Kaul, V. (1984). Conservation of plant resources in aquatic ecosystems with reference to some aquatic habitats of Kashmir – In Conservation of Tropical Plant Resources (ed. Jain & Mehra), BSI, Calcutta, 118-131.

Khan, A.H. (1958). The essential oil bearing plants of Pakistan – Part II, Pakistan J. For., 8 (4): 342-373.

Khanna, P. (1965). Morphological and embryological studies in Nymphaeaceae – II: Brasenia schreberei Gmel. and *Nelumbo nucifera* Gaertn., Austr. J. Bot., 13 (3): 379-87.

Khoshoo, T.N. (1971). Taxonomy of cultivated plants: Principles, Procedures and Prospects, Sci. & Cult, 37 (7): 313-315.

Khoshoo, T.N. (1980). Taxonomy of cultivated plants: Some basic considerations, in Modern Trends in Plant Taxonomy, Glimpses in Plant Research –V (ed. P.K.K. Nair), 306-314.

Kunitomo, J., Yamamoto, C. and Otsuki, T. (1964). Alkaloids of *Nelumbo nucifera* – VI : Alkaloids of "genshihasu", Yakugaku Zasshi, 84 (11): 1141-1142.

Lakhanpal, R.N. (1955). On the occurrence of *Nelumbium* in the tertiary of Assam, J. Indian Bot. Soc., 34 (3): 222-224.

Lavania, G.S., Paliwal, S.C. and Gopal, B. (1990). Aquatic vegetation of the Indian subcontinent, pp. 29-78, in Ecology and Management of Aquatic Vegetation In the Indian Subcontinent (ed. Brij Gopal), Kluwer Academic Publishers, Netherlands, 29-78.

Lawrence, G.H.M. (1963). The Taxonomy of cultivated plants, in Vistas in Botany-II (ed. W.B. Turrill), Oxford & IBH Publishing Co., New Delhi, 199-214.

Li, H.L. (1955). Classification and phylogeny of the Nymphaeaceae and allied families, Am. Midland Nat., 54 (1): 33-41.

Li, J.K. and Huang, S.Q. (2009). Flower thermoregulation facilitates fertilization in Asian sacred Lotus, Annals of Botany, 103 (7): 1159-1163.

Li, L.J. and Wen, Z.Y. (1998). J. Nanjing Agril. Univ., 21 : 113-115.

Liang Jung Li and Zhao Yao Wen (1998). J Nanjing Agril. Univ., 21: pp. 113-115.

Lin, Z., Zhang, Cheng, Cao, D., Damaris, R.N. and Yang, P. (2019). The latest studies on lotus (*Nelumbo nucifera*) – An emerging horticultural model plant, Int. Jour. Mol. Sci., 20 (15): 3680.

Lindley, J. (1836). A Natural System of Botany, London, 1.

Marmur, A. (2004). A botanical perspective on the identity of 'Soma' (*Nelumbo nucifera* Gaertn.) based on Scriptural and Iconographic Records, Econ. Bot., 58 (Supp): 147-173.

Mc Donald, A. (2004) A bootanical perspective on the identity of 'Soma' (*Nelumbo nucifera* Gaertn.) based on Scriptural and lconogrphic Records, Econ. Bot., 58 (Supplement): pp. 147-173.

Meyer, W.C. (1930). Dormancy and growth studies of the American Lotus, *Nelumbo lutea*, Plant Physiol., 5 : 225-234.

Mitra, R. and Kapoor, L.D. (1976). Kamala – The National Flower of India: Its Ancient History and Uses in Indian Medicine, Ind. J. Hist. Sci., 11 (2): 125-132.

Mohan Ram, H.Y. (1991). Biology of aquatic flowering plants, Proc. Indian Nat. Sci. Acad, 57 (B): 95-108.

Mukherjee, P. (2002). Quality control Herbal drugs: an approach to evaluation of botanicals, Business Horizons, New Delhi, 1.

Mukherjee, P.K., Mukherjee, D., Maji, A.K., Rai, S. and Heinrich, M. (2010). The sacred lotus (*Nelumbo nucifera*) phytochemical and therapeutic profile, J Pharm Pharmacol, 61 (4): 407-422.

Nadkarni, A.K. (1954). *Nelumbium speciosum* Willd. K.M. Nadkarni's Indian Materia Medica, Revised and enlarged ed., Popular Book Depot, Bombay, 844-845.

Nagarajun, S., Nair, A.G.R., Ramakrishnan, S. and Subramaniam, S.S. (1966). Chemical examination of the flowers of *Nelumbium speciosum* Willd., Curr. Sci., 35 (7): 176-177.

Nakaoiki, T., Torita, N., Nagata, Y. and Oguri, H. (1961). Medicinal resources XIX. Flavonoid of the leaves of *Nelumbo nucifera*, Cosmos bipinnatus and Foeniculum vulgare, Yakugaku Zasshi, 81 : 1158-1159.

Nguyen, V.Q. (1998). Report on the study tour to Japan on pickling of Asian vegetables and attendance at an international symposium on vegetable quality in Seoul, Korea 1997, Gosford, Horticultural Research and Advisory Station, 1-96.

Nohara, S. and Tsuchiya (1990). Effects of water level fluctuation on the growth of *Nelumbo nucifera* Gaertn. in lake Kasumigaura, Japan, Ecological Research, 5 : 237-252.

Ohga, I. (1926). A comparison of the life activity of century-old and recently harvested Indian lotus fruits, Amer. J. Bot., 13 (10): 760-765.

Ohga, I. (1926a). A germination experiment on the fruit of Indian lotus collected by Sir Hans Sloane which was preserved at the British Museum, Bot. Mag, 40 (471): 153-158.

Ohga, I. (1926b). A report on the longevity of the fruit of *Nelumbium*, Amer. J. Bot., 64 (762): 154-157.

Ong, H.C. (1996). *Nelumbo nucifera* Gaertner. In: Plant Resources of South-East Asia No. 9: Plants yielding non-seed carbohydrates (Ed; Flach, M. and F. Rumawas), Wageningen, The Netherlands, Pudoc Scientific Publishers, 131-133.

Perry, F. (1989). Collin's Guide to Waterlilies and other aquatic plants, pp. 160, London, 98-105.

Porterfield, W.M. (1941). Lotus plants for ornaments, food and medicine, J. New York Bot. Gard., 42 : 280-286.

POWO (2021). "Plants of the World online, Facilitated by the Royal Botanic Gardens, Kew (http://www.plantsoftheworldonline.org)

Puri, G.S. (1950). Note on the fossil leaf of *Nelumbo nucifera* Gaertn., from the pleistocene (Karewa) deposits of Kashmir with a review of the history of the Nymphaeceae, Indian Forest., 76 : 343-346.

Rahman, W., Ilyas, M. and Khan, A.W. (1962). Flower pigments: flavonoidsfromNelumbonuciferaGaertn.,Naturwissenschaften, 49 : 327-1.

Rheede, H.A. Van (1678-1693). Horti Indici Malabarici. 12 Vols., Amsterdam, Netherlands.

Roxburgh, W. (1874). Flora Indica: On description of Indian plants, II, Serampore, 1.

Sandhya, B., Thomas, S., Isabel, W. and Shenbagarathai, R. (2006). Ethnomedicinal plants used by the Valaiyan Community of Pirammalai Hills (Reserved Forest), Tamil Nadu, India - A pilot study, African Journal of Traditional, Complimentary and Alternative Medicines, 3 (1): 101-114.

Schneider, E.L. and Buchanan, J.D. (1980). Morphological studies of the Nymphaeaceae – XI, The floral biology of *Nelumbo pentapetala*, Amer. J. Bot., 67 (2): 182-193.

Schneider, E.L. and Kodono, Y. (1984). Floral Biology, development and Structure of Euryale ferox, Amer. J. Bot., 71-131.

Sculthorpe, C.D. (1971). The biology of aquatic vascular plants 2nd edition, Rep. 1985. Bishen Singh Mahendra Pal Singh, Dehradun, 610.

Seymour, R.S. and Motel, P.S. (1996). Thermoregulating lotus flowers, Nature, 383 (6598): 305.

Sharma, S.C. and Goel, A.K. (1997). FTR on the Project – Introduction, Multiplication, Conservation and Documentation of *Nelumbo* (Lotus) and Euryale spp./cvs., Min. of Environ. & For., New Delhi, 1.

Sharma, S.C. and Goel, A.K. (2000). Environmental degradation and ex-situ conservation of Nelumbo nucifera, in Environmental Stress: Indication, Mitigation and Eco-conservation (Eds. M. Yunus, N. Singh & L. J. de Kok), Kluwer Academic Publishers, Netherlands, 405-410.

Sharma, S.C. and Goel, A.K. (2001). Conservation of Lotus (*Nelumbo nucifera*), Ind. Hort, 46 (1): 18-22.

Sharma, S.C. and Goel, A.K. (2006). Horto-Taxonomy of Ornamentals: Case Study of B*ougainvillea*, J. Econ. Tax. Bot., 30 (1): 71-78.

Sharma, S.C., Goel, A.K. and Roy, R.K. (1998) NBRI Newsletter, 25 : 34-35.

Shen-Miller, J., Mudgett, M.B., Schopf, J.W., Clarke, S. and Berger, R. (1995). Exceptional seed longevity and robust growth: ancient sacred lotus from China, Amer. J. Bot., 82 (11): 1367-1380.

Shen-Miller, J., Schopf, J.W., Harbottle, G.,Cao, R.J., Ouyang, S., Zhou,K.S., Southon, J.R. and Liu, G.H. (2002). Long living lotus: Germination and soil irradiation of centuries-old fruits and cultivation, growth, and phenotypic abnormalities of offsprings, Amer. J. Bot., 89 (2): 236-247.

Shivalingappa, H., Satyanarayan, N.D., Purohit, M.G., Sharanabasappa, A. and Patil, S.B. (2002). Effect of ethanol extract of Rivea hypocrateriformis on the estrous cycle of the rat, J Ethnopharmacol, 82 : 11-17.

Sinha, S., Mukherjee, P.K., Mukherjee, K., Pal, M., Mandal, S.C. and Saha, B.P. (2000). Evaluation of antipyretic potential of *Nelumbo nucifera* stalk extract, Phytother. Res, 14 (4): 272-274.

Slocum, P.D. (1985). Propagating water Lilies and aquatics, Brookleyn Botanic Garden Record, 41 : 25-28.

Slocum, P.D. (2005). Waterlilies and Lotuses: species, cultivars and new hybrids, Timber Press, USA, 236.

Slocum, P.D. and Robinson, P. (1996). Water gardening, Water lilies and lotuses, Timber Press, USA, 434.

Stafleu, F.A. and Cowan, R.S. (1988). Taxonomic literature, 7 vols.– Bohn, Scheltema and Holkema, Utrecht, 1.

Stefferud, A. (1959). The Wonders of Seeds, MacMillan & Co. Ltd., London, 112.

Strauss, J.F. III, Schuler, L.A. and Tanaka, T. (1981). Cholesterol metabolism by ovarian tissue, Advances in Lipid Research, 18 : 99.

Subramanyam, K. (1962) Aquatic Angiosperms, p.190, CSIR, New Delhi, 8.

Suijian, Q. and Delong, Z. (2013). Lotus seed epicarp extract as potential antioxidant and anti-obesity additive in Chinese Cantonese Sausage , Meat Science, 93 : 257-262.

Swarup, V. (1989). Lotus, In Bose & Yadav (ed.), Commercial Flowers, 7-13.

Takhtajan, A. (1959). Die Evolution der Angiospermen, Jena, Germany, 1.

Thiagrajan, G. (1992). In pursuit of better image: Survey of the environment, The Hindu, 143-145.

Thorne, R.F. (1983). Proposed new realignments in angiosperms, Nord. J. Bot, 3 : 85-117.

Toyoda, K. (1960). The ratio of chlorophyll a and b in the plumule of *Nelumbo nucifera*, Bot Mag, 73 : 398-403.

Trivedi, A., Mishra, P. and Shrivastava, M.P. (1980). Pharmacognostical investigation of the starches from the rhizomes of *Canna coccinea* Mill. and *Nelumbo nucifera* Gaertn, East Pharm, 23 : 185-186.

Trivedi, A., Shrivastava, M.P. and Mishra, P. (1982). Studies on binding and disintegrating properties of *Nelumbo* starch, East Pharm, 25 : 199-202.

Vajpayee, P., Sharma, S.C., Tripathi, R.D., Rai, U.N. and Yunus, M. (1999). Bioaccumulation of Chromium and Toxicity to Photosynthetic Pigments, Nitrate Reductase Activity and Protein Content of *Nelumbo nucifera* Gaertn., Chemosphere, Elsevier Science Ltd, 39 (12): 2159-2169.

Varshney, C.K. and Singh, K.P. (1976). A Survey of aquatic weed problem in India, in: Varshney & Rzoska (eds.), Aquatic Weeds in South-East Asia, Junk, The Haque, 31-42.

Wagner, T., Neinhuis, C. and Barthlott, W. (1996). Wettability and contaminability of insect wings as a function of their surface sculpture, Acta Zoologica , 77 : 213-225.

Ward, D.B. (1977). *Nelumbo lutea*, The correct name for the American Lotus, Taxon, 26 (2/3): 227-234.

Watt, G. (1972). *Nelumbium speciosum* Willd. In: A Dictionary of the economic products of India, Periodical Experts, Delhi, 5 (V): 343-345.

Willdenow, C.L. (1800). Caroli a Linne Species Plantarum , G.C. Nauk. Berlin, 1.

Wilson, R.F. (1938). Horticultural Colour Chart I & II, British Colour Council in Collaboration with Royal Horticultural Society, London, 200.

Wu, J., Zheng, Y., Chen, T., Yi, J., Qin, L., Rahman, K. and Lin, W. (2007). Evaluation of quality of lotus seeds *Nelumbo nucifera* Gaertn. From out space mutation, Food Chem, 105 : 540-547.

Xia, X. (2012). Effect of germination on chemical and functional properties of lotus seeds, Food Sci, 33 : 91-98.

Yamaguchi, M. (1990). Asian vegetables. Advances in new crops. Ed.: J. Jamick and J.E. Simon. Portland, Oregon, Timber Press, USA, 387-390.

Yun, J., Tzi, B.N.G., Zhaokum, L.I.U.,Changrong, W., Ning, Li, Wentao, Q.I.A.O. and Fang, L.I.U.A. (2011). Immunoregulatory and anti-HIV- 1 enzyme activities of antioxidant components from Lotus (*Nelumbo nucifera* Gaertn.) rhizome, Biosci. Rep, 31 : 381-390.

Zhang, Y, Lu, X, Zheng, S., Huang, Y., Guo, Z., Zheng, Y., Tian, Y. and Zheng, B. (2015). Nutritional composition, physiological functions and processing of lotus (*Nelumbo nucifera*) seeds- A review, Phytochem. Rev, 14 (3): 321-334.

Pink double lotus flowers

***Nelumbo* sp. having white coloured flower**